Patrick Moore's
Practical Astronomy series

Springer
London
Berlin
Heidelberg
New York
Barcelona
Hong Kong
Milan
Paris
Singapore
Tokyo

Other titles in this series

Solar Observing Techniques

Chris Kitchin

With 119 Figures
(including 25 Colour Plates)

Springer

British Library Cataloguing in Publication Data
Kitchin, Christopher R. (Christopher Robert), 1947–
Solar observing techniques. – (Patrick Moore's practical
astronomy series)
1. Sun – Observations
I. Title
523.7
ISBN 185233035X

Library of Congress Cataloging-in-Publication Data
Kitchin, C. R. (Christopher R.)
Solar observing techniques / Chris Kitchin.
    p. cm. – (Patrick Moore's practical astronomy series, ISSN 1431-9756)
Includes bibliographical references and index.
ISBN 1-85233-035-X (acid-free paper)
1. Sun–Observers' manuals. 2. Sun–Amateurs' manuals. I. Title.
II. Series.
QB521.K65 2001
523.7–dc21   2001017009

Patrick Moore's Practical Astronomy Series ISSN 1431-9756
ISBN 1-85233-035-X Springer-Verlag London Berlin Heidelberg
a member of BertelsmannSpringer Science+Business Media GmbH
http://www.springer.co.uk

© Springer-Verlag London Limited 2002
Printed in Singapore

Typeset by EXPO Holdings, Malaysia
Printed and bound by Kyodo Printing Co. (S'pore) Pte. Ltd., Singapore
58/3830-543210 Printed on acid-free paper SPIN 10676081

For Christine

Preface

The purpose of this book is to introduce the reader to the numerous safe methods of observing the Sun and solar eclipses, and to suggest objects and features to observe and observing programmes to follow. So much energy comes from the Sun that by failing to observe safe working practices it is possible to damage your eyes or equipment. The care that is needed is emphasised throughout the book. Always make sure that you have read the whole of a section or chapter before starting any observational work. However in warning when care is needed in observing, there is a danger of scaring people off observing the Sun altogether. Let me emphasise therefore that observing the Sun can be done in complete safely, providing that the precautions discussed in the book are followed. The Sun then provides one of the most interesting objects in the sky for an astronomer to study at all times, and during a total solar eclipse becomes uniquely fascinating to both astronomers and the general public alike.

So take heed of the warnings given here but do not let them stop you trying out the safe observing methods. I wish you clear sunny skies and many hours of fun.

Chris Kitchin
Hertford, 2001

Acknowledgements

I would like to thank Dr Ralph Chou for his help in supplying details of filters and of sources of further information on them. I would also like to thank the University of Hertfordshire for the loan of some of the equipment illustrated in here and used to obtain some of the solar images.

Contents

The Alert Symbol

**CAUTION
REQUIRED**

The alert symbol is used throughout this book to warn the reader of potentially dangerous procedures. You should always make sure that you have read the whole of the relevant warnings, sections or chapters before trying anything out for yourself, but that advice is doubly necessary for anything with this symbol attached.

Warning

**CAUTION
REQUIRED**

So much energy comes from the Sun that it is easy to cause blindness or other damage to your eyes or to your telescope if suitable preventative measures are not undertaken first. You should note that the retina of the eye does not contain pain receptors. You will therefore **not** feel pain when damage is occurring, and often the damage does not become apparent until several hours after it has occurred.

Much of the Sun's energy is in the infrared part of the spectrum. Simply reducing the intensity of the visible light is therefore **not** sufficient; the infrared component must be eliminated as well. **Never** look directly at the Sun, even when it is low on the horizon, and seems dim enough to be safe. A professionally produced filter is essential even for naked eye work. Do not use any of the following as substitute filters:

- smoked glass,
- exposed black and white film[1]

---

[1] A double layer of silver-based black and white photographic film can be used as a safe solar filter. The film must be exposed for a minute or so to direct solar light, and then fully developed. A negative containing images is **not** safe. It is essential that the emulsion is silver-based. Dye-based black and white films, and colour films are **not** safe. Colour film for example can have 50% transmission in the near infrared, compared with the safe limit of 0.027%. Unless you are knowledgeable about photographic emulsions and can be certain of using a silver-based product then do not be tempted to try this approach.

- CDs
- space blankets
- aluminised helium balloons
- aluminised potato crisp packets
- floppy disks
- crossed Polaroid filters
- smoked plastic
- sunglasses
- mirrors
- X-ray photographs.

These items are **not** safe and may let through the infrared part of the Sun's energy. For binocular and telescopic work even greater precautions are required, as detailed in the following sections and chapters of this book. Do **not** point a pair of binoculars or a telescope anywhere near to the Sun in the sky until you have completely assimilated the precautions and techniques required and which are discussed below.

Even with the precautions described in this book, if you have extra-sensitive eyes, or suffer from other eye problems, consult your optician before undertaking any solar observing.

While the information in this book is given in good faith, neither the author nor the publisher accept any responsibility for injury to persons or for damage to equipment resulting from observing or attempting to observe the Sun.

The Sun radiates energy at all wavelengths from the shortest of gamma rays to the longest radio waves. Our atmosphere, however, shields us from most of this emission by absorbing it before it reaches the surface of the Earth. The atmosphere is more or less transparent over two wavelength regions; the optical[2] and radio regions. These two transparent regions are often called the atmosphere's "windows" (Fig. 0.1).

Radiation at any wavelength can be dangerous if its intensity is high enough. The solar radio radiation, however, is too weak to cause any problems. Thus we are concerned only with that which comes through the optical window.

It is usually stated that the ultraviolet limit of the optical window is about 320 nm.[3] However this is

---

[2] The optical region comprises near ultraviolet, visible and near infrared radiation.

[3] Abbreviation for nanometre. 1 nm $= 10^{-9}$ m $= 0.004$ millionths of an inch.

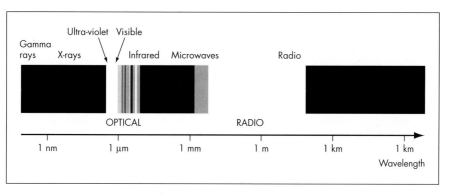

misleading. The atmosphere is not completely opaque; it absorbs only a proportion of the radiation. The proportion absorbed varies with the wavelength and the amount of the atmosphere along the line of sight. At 320 nm, when the Sun is at the zenith, about 50% of the radiation will still reach the surface of the Earth at sea level. That will fall to about 30% when the Sun's altitude above the horizon is 45°, and to about 10% for an altitude of 20°. Conversely, at a height of 5000 m (3 miles) above sea level about 75% of the radiation will still get through the atmosphere with the Sun overhead. Not until the wavelength is as short as 290 nm is the atmosphere effectively completely opaque. Radiation below 315 nm is termed UV-B,[4] and is particularly damaging to biological materials.

You should note that if observing from a high altitude site, from an aeroplane, or from high latitudes where the Earth's ozone layer is depleted,[5] then the intensity of the ultraviolet radiation will be increased and shorter wavelength ultraviolet radiation may penetrate through at significant intensities. Extra

---

[4] UV-A radiation lies between 315 nm and 380 nm.

[5] This is due to the ozone holes which are regions of the Earth's atmosphere where the ozone (an oxygen molecule containing three oxygen atoms) is depleted compared with its normal values. The depletion arises through the action of CFCs (chlorofluorocarbons) and the exhaust gases from high flying jet planes. The holes are found around the north and south poles of the Earth, and can extend (at the time of writing) at least down to latitudes of about 45°. The holes vary in depth, size and position with the seasons, normally being worst in winter and early spring. The reduced ozone content of the atmosphere allows shorter wavelength ultraviolet radiation than normal to reach the Earth's surface and also

(footnote 5 continues overleaf)

precautions, such as the use of an ultraviolet blocking filter,[6] should therefore be employed when observing in these situations. Alternatively welder's filters (Chapter 6) should be used since these have an intrinsically high ultraviolet absorption because of the intense levels of that radiation from welding arcs.

The eye's retina responds to radiation from about 380 nm[7] (deep violet) to 700 nm (deep red). The radiation from about 290 nm to 380 nm does not reach the retina, but is absorbed in the preceding parts of the eye. That absorption can damage the outer layers of the eye leading to an increased risk of cataracts and to enhanced ageing of the eye. The damaging effect increases rapidly as the wavelength gets shorter.

The eye has evolved to utilise the radiation from 380 nm to 700 nm, and this is therefore not intrinsically damaging. However, at very high intensities, such as those that we receive directly from the Sun, even these wavelengths can damage the retina. As previously noted (Warning, above), you will not get the usual physiological notification of damage occurring (pain) because of the lack of pain receptors in the retina. The damage occurs through reducing the sensitivity of the receptor cells of the retina (the rods and cones), or at extreme levels of illumination, through the heating effect (cooking!). The former effect can be temporary, but is more often permanent, while the latter effect always results in permanent damage. The damaged areas are small, but will normally be around the centre of the eye's field of view, and thus be most incapacitating.

Infrared radiation out to wavelengths of about 1.4 μm (1400 nm) is transmitted through the eye, though the retina does not respond to wavelengths longer than

---

(footnote 5 continued)
results in higher than normal intensities for the longer wavelength ultraviolet radiation. As well as the precautions outlined here for actually observing the Sun, it is therefore also necessary to take more general precautions against sunburn and skin cancer through the use of high factor sun blocking creams, suitable sunglasses, etc.

[6] Obtainable as an attachment for camera lenses, but check with the manufacturer what wavelengths are blocked.

[7] People who have had the natural lens of the eye removed because of cataracts or other injury can sometimes see to shorter wavelengths than 380 nm. Especial care is needed in such cases because the retina can additionally be damaged by the short-wave radiation which is now able to reach it.

700 nm. Nearly as much of the Sun's energy (37%), however, lies in the region from 700 nm to 1.4 μm as lies in the visual region (39%). It is thus particularly important to ensure that this part of the spectrum is eliminated by any method used to observe the Sun, since not even the appearance of excessive brightness will be available to warn of potential damage.

These rather dire warnings may have left you wondering about observing the Sun at all. Let us therefore end this section on a positive note. Provided that you take the precautions detailed in the following sections and chapters of this book, then observing the Sun can be safe, enjoyable and instructive. Like most potentially dangerous activities, the danger only arises through inexperienced, or foolish, behaviour. With appropriate care and a sensible approach, there are no more difficulties in solar observing than those in any other form of astronomy.

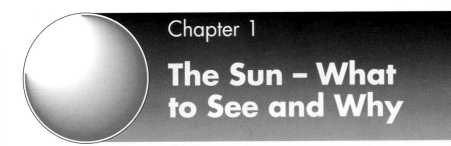

Chapter 1

The Sun – What to See and Why

Beginnings

The Sun is often neglected by amateur astronomers except to wish that it set earlier or rose later so that they can get on with "real" astronomy. This is a great mistake since anyone with a small telescope – and you do not need huge instruments to study the Sun – can easily observe the most fascinating phenomena on the one object in the sky that directly influences our everyday lives. Furthermore the Sun is the one star that can easily be observed in great detail. Solar observing does require a small amount of specialised equipment, but the expenditure compared with the cost of the telescope is small. Solar observing also has the merit that it is undertaken on a nice warm sunny day, which has its advantages over working through the icy blasts of a mid-winter night!

The following chapters of this book go into details of what is needed for solar observing and how to undertake it. Here, let us begin by reviewing what there is to see on the Sun and what is its significance. Note, however, the warning at the start of this book and do not be tempted to start observing yourself, until you have read all the relevant later sections.

Sunspots

Sunspots will be encountered again in Chapter 3, and they are by far the most prominent and well known of

Figure 1.1. Sunspots over the whole Sun.

the solar phenomena (Fig. 1.1). They are fascinating individually and in toto, for it is their comings and goings which are the basis of the Sunspot cycle. The latter has direct effects upon the Earth and our lives, ranging from variations in the average temperature (the Earth is slightly hotter when there are many sunspots) through the production of the radioactive isotope, carbon-14, used by archaeologists for dating their finds, to the degradation of the solar panels powering spacecraft and the appearances of the aurorae.

Sunspots are cooler regions of the Sun's surface. Their centres have temperatures around 4200 K compared with 5800 K for the normal solar surface. The material within a sunspot therefore has a surface brightness only about a quarter that of the photosphere. By contrast with the solar photosphere, the sunspots appear to be very dark. However if we were able to see sunspots by themselves, they would be very bright objects; a sunspot just 10″ across[8] would emit several times the energy coming from the full moon.

---

[8] Angles smaller than one degree are measured in minutes of arc (symbol ′) and seconds of arc (symbol ″).

$$1° = 60' \text{ and } 1' = 60''.$$

One second of arc is thus 1/3600 of a degree; the size of a 10 mm circle at a distance of 2 km (a 1-inch circle at a

Figure 1.2. The magnetic field near a sunspot.

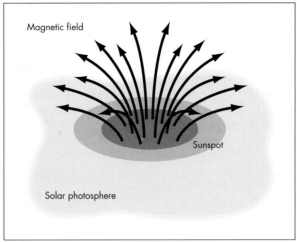

The reason for the lower temperature in sunspots is because of the presence of very intense magnetic fields within them. The general solar magnetic field varies considerably but is typically two or three times stronger than the Earth's magnetic field. However, inside sunspots the field strength rises to between 2000 and 10,000 times that of the Earth's field. Now in the outer layers of the Sun, the energy generated at its centre is largely carried outwards by convection currents (Fig. 1.12). Within sunspots, the magnetic field is coming out of the solar surface (Fig. 1.2), and it suppresses that convection below the spot. With the convection suppressed, or at least reduced in its efficiency, less energy reaches the surface of the Sun over the region covered by the magnetic field. That region therefore cools down, and appears to us as a sunspot.

A simple sunspot is more or less circular and has a dark centre termed the umbra, and a lighter surrounding region which has a radial structure (Figs. 1.3 and 1.4) called the penumbra. Sunspots usually appear in pairs, with one as a magnetic north pole and the other as a magnetic south pole. Sometimes the second spot

---

(footnote 8 continued)
distance of 3 miles). An object one second of arc across is about the smallest thing that can be seen clearly through any telescope on the Earth, because of the blurring effect of the Earth's atmosphere. An object 2–3′ (100,000 km on the Sun) across could theoretically just about be seen directly on the Sun with the unaided eye, but remember the warning about **not** observing the Sun directly.

Figure 1.3. Typical small sunspots (photograph reproduced by permission of the Royal Astronomical Society).

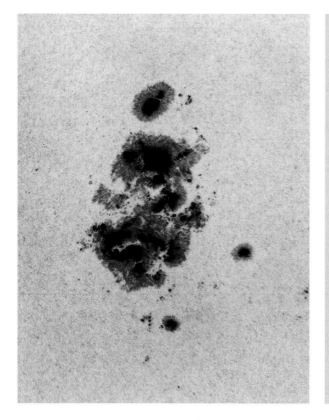

Figure 1.4. A sunspot group, with individual spots showing their structure (photograph reproduced by permission of the Royal Astronomical Society).

does not appear, but the magnetic fields are still there and can be detected either directly or through the presence of plages and prominences (see later in this chapter). The magnetic field loops out of the solar surface through one spot and returns via the other. The smallest spots comprise just the umbra without the penumbra and are called pores. Larger spots are often complex (Fig. 1.4), and made up from many simple and merged individual spots. They are then called sunspot groups, and they may also have highly complex and intertwined magnetic fields. The sizes of individual spots range from a few thousand to a hundred thousand kilometres (1000 to 50,000 miles), and large groups can extend to over 150,000 km (100,000 miles). The lifetimes for the visible spots range from less than a day to several months. A useful rule of thumb is that the lifetime of a spot in days is about an eighth of its corrected area measured in millionths of a solar hemisphere,[9] when at its maximum size.

Spots first appear on the solar surface as pores. In most cases these then expand and develop a penumbra. Soon afterwards the second spot of the pair usually appears. This takes a few days, and for a typical spot pair, the process will then reverse and rather more slowly the spots will disappear. With large groups the initial development is rapidly followed by the appearance of many more small spots and the merging of large and small spots to form complex shapes. The peak development of a sunspot group is reached after about a quarter of the whole lifetime of the group.

---

[9] Sunspot areas are often measured in units of one millionth of the area of the solar hemisphere. To do this, the observed projected area must be corrected for foreshortening if the spot is not at the centre of the solar disk. If the spot is x% of the way from the centre of the solar disk to its edge, and if the projected spot area is measured in square seconds of arc, then the corrected spot area in millionths of a hemisphere is:

$$\text{Corrected Area} = \frac{\text{Projected Area}}{5(1 - (\frac{x}{100})^2)^{\frac{1}{2}}}$$

where the corrected area is in millionths of a hemisphere and the projected area is in square seconds of arc.

One millionth of a hemisphere is equal to 5 square seconds of arc. The lifetime in days may therefore also be estimated as about 2.5% of the maximum corrected area measured in square seconds of arc.

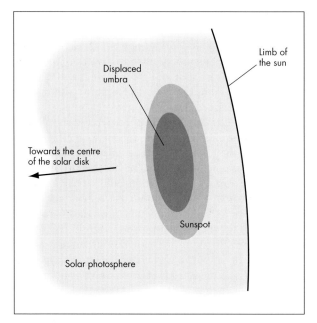

Figure 1.5. The Wilson effect.

The apparent surface of the umbra (see footnote 10 below for what is meant by a surface in a gaseous medium) is about 500 km (300 miles) lower than that of the photosphere. This can lead to a phenomenon first observed in 1769 and named after its discoverer: the Wilson effect. This effect is that when a spot is seen near the solar limb, the umbra is no longer central within the penumbra, but is displaced towards the centre of the solar disk (Fig. 1.5). Since spots are rarely perfectly symmetrical, the effect is not always to be seen, and may even be reversed for some spots.

There are many other features associated with sunspots as outlined later. The whole set of features; sunspots, faculae, plages, flares, etc., is called an active region.

Sunspot Cycle

Individual sunspots come and go with lifetimes ranging from a few hours to many weeks. The appearance of the Sun is therefore always changing as new spots develop and old ones disappear. But if the overall appearance of the Sun is recorded over several years, then it is found that there is a pattern to the changes. At some times there are many more sunspots to be seen than at others, and the interval between successive sunspot maxima or minima is about 11 years. This variation in the number of sunspots is the well-known sunspot cycle. Solar cycles are numbered and number 23 started in late 1997/early 1998.

There are several ways of measuring the sunspot activity, and for high levels of accuracy the area covered by the spots is the best criterion. However for many purposes, a measure which is much easier to obtain is quite sufficient, and this is the Zurich or Wolf relative sunspot number. It is obtained by simply counting the number of individual spots and sunspot groups to be seen on the Sun; the individual areas are not measured. The sunspot number, R, is then given by

$$R = k(10g + s)$$

where g is the number of groups (an isolated spot counts as a group), s is the number of individual spots (including those in groups), and k is a correction factor, usually close to 1, to allow for the efficiency of the observer and his/her equipment. It has to be determined by finding the spot number and comparing the observer's value with the official value at the same instant over a period of several months.

Normally the sunspot number is averaged over a 27-day period so that the effects of solar rotation are smoothed out. The variation in R since 1900 is shown in Fig. 1.6. General levels of solar activity can also be monitored by the solar short-wave radio emission (Chapter 9).

Figure 1.6 shows several of the features of the sunspot cycle:

- the rise from minimum to maximum is somewhat faster than the fall from maximum back to minimum;
- the cycle length averages 11 years, but can vary considerably from that. Cycles as long as 14 years and as short as 8 years have been recorded;
- the sunspot numbers at maximum and minimum can vary considerably from one cycle to the next.

Several features of the sunspot cycle, however, are not apparent from Fig. 1.6. The first of these is the latitudes at which spots are seen. At the start of a new cycle, the spots appear at solar latitudes of 30° to 40° north or

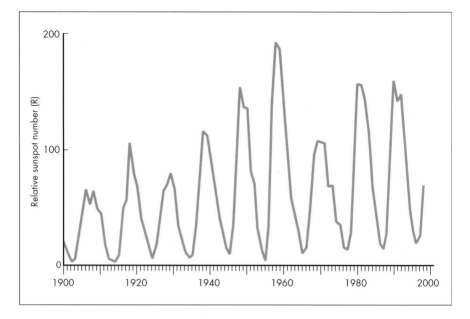

Figure 1.6. Sunspot numbers (R) since 1900.

south. As the cycle proceeds, spots appear at lower and lower latitudes, until by the end of the cycle, they are appearing close to the equator. This drift is called Spörer's law and is graphically shown on the "Butterfly diagram" (Fig. 1.7). The butterfly diagram is a plot against time of the latitudes of sunspots. The drift and the change in sunspot numbers combine to give the appearance of a butterfly's wings.

Also not apparent from Fig. 1.6 is that the sunspot cycle is actually 22 years long, not 11. This only becomes clear when the magnetic fields are taken into account. During one 11-year cycle the leading spot in one hemisphere (in the sense of the Sun's rotation) will always be of the same magnetic polarity. The trailing spot in that same hemisphere will be of the opposite polarity. In the other hemisphere throughout that same sunspot cycle, the polarities will be reversed. Through-out the next 11-year cycle the whole pattern is reversed (Fig. 1.8). Thus it takes two of the basic cycles for the magnetic patterns to recur.

Finally, Fig. 1.6 does not go far enough back in time to show the remaining peculiarity of the sunspot cycle, which is that it is not always present. For periods ranging from a few decades to two or three centuries, sunspots will sometimes disappear entirely from the Sun. The last such period was from 1645 to 1715, and is known as the Maunder minimum. It is probably not a

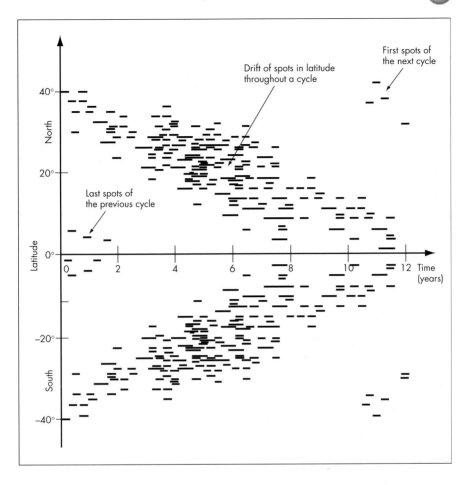

Figure 1.7. The butterfly diagram (schematic) showing the variation in numbers and latitude of sunspots throughout a sunspot cycle.

coincidence that it happened at the same time that a "mini ice age" occurred on the Earth, since the Sun seems to get slightly more luminous, the more sunspots that there are present on it. There is evidence (from carbon-14 abundances which are related to solar cosmic ray intensities) of previous periods when sunspots vanished, such as the Spörer minimum from 1450 to 1550.

Limb Darkening, Granulation and Faculae

Sunspots are easily seen, indeed, they are sometimes visible to the unaided eye (although for such observations

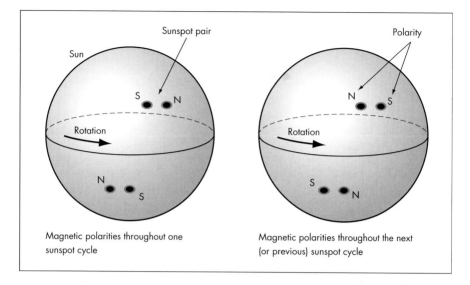

Magnetic polarities throughout one sunspot cycle

Magnetic polarities throughout the next (or previous) sunspot cycle

filters, etc., still need to be used – see Chapter 6). Several other features of the Sun can be seen in white light, but they are rather more elusive than the spots. Techniques for observing these features are discussed in Chapter 3.

Figure 1.8. Magnetic polarities in sunspots.

Limb Darkening

At first sight and ignoring sunspots, the solar disk often seems to be featureless. Careful inspection, however, will show that the surface brightness fades towards the edges or limbs of the Sun (Fig. 1.9). This is the phenomenon called limb darkening and it arises because the Sun is not a solid body. In any object composed of a semi-transparent medium, we can receive light from a whole range of points along the line of sight, not just from a single surface layer as would be the case for a solid opaque object like the Earth or the Moon. None the less, when we look at such an object there does appear to be a definite surface. That layer, called the photosphere for the Sun, corresponds to the depth at which about a third[10] of the radiation escapes

---

[10] A more accurate figure is 36.8%. The visible surface occurs at an optical depth of 1. The optical depth (τ) is the exponent in the equation

$$I_{\text{Observed}} = I_{\text{Layer}} \times e^{-\tau}$$

where I_{Observed} is the energy reaching the observer and I_{Layer} is the energy present within the layer.

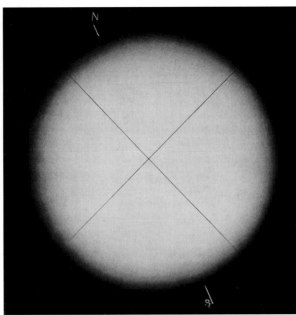

Figure 1.9. Limb darkening (photograph reproduced by permission of the Royal Astronomical Society).

without being absorbed in the higher layers. We thus receive about a third of the energy which is actually present at the apparently visible surface of the Sun, and two-thirds is absorbed in the layers of the Sun above the visible surface. We also, however, receive about 60% of the energy from a layer 30 or 40 km (20 miles) above the visible surface, 90% from a layer 100 km (60 miles) above the visible surface, 14% from a layer 30 km (20 miles) below the visible surface, and so on.

Where the visible surface of the Sun occurs is thus determined by the absorption in the layers above that surface. When we look near the edge of the Sun, the line of sight is passing at a shallow angle through the upper layers. The point at which two-thirds of the energy will be absorbed before it escapes is thus reached within a layer which is higher in the Sun's atmosphere than when we look towards the centre of the solar disk and the line of sight is passing vertically through the layers (Fig. 1.10).

But the temperature in the Sun's outer layers is decreasing over the regions that we are considering here (Fig. 1.11). The layer corresponding to the visible surface at the centre of the Sun's disk is lower down in the solar atmosphere and so at a higher temperature than the layer corresponding to the visible surface near the edge of the disk. The centre of the disk thus appears brighter than the edges, and so we have limb darkening.

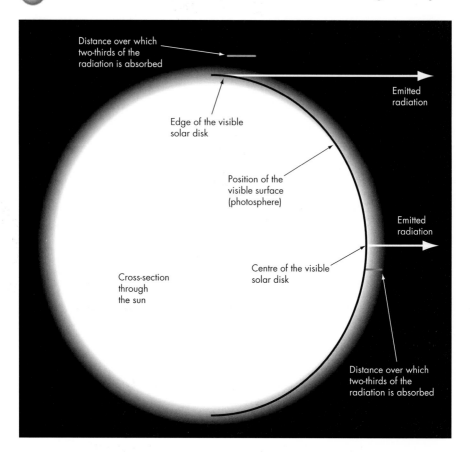

Distance over which
two-thirds of the
radiation is absorbed

Emitted
radiation

Edge of the visible
solar disk

Position of the
visible surface
(photosphere)

Emitted
radiation

Cross-section
through
the sun

Centre of the visible
solar disk

Distance over which
two-thirds of the
radiation is absorbed

At visual wavelengths the limb intensity is about 40% of that at the centre of the solar disk.

Figure 1.10. The visible solar surface.

Granulation

Under excellent observing conditions and using an instrumental set-up that gives good contrast, it is possible to see that the whole photosphere has a mottled appearance (Figs. 1.3 and 1.4, for example, outside the area covered by the sunspots). The use of a CCD detector (Chapter 4) may help in showing this pattern since its peak sensitivity is in the near infrared part of the spectrum where the scattered solar light in the Earth's atmosphere is of reduced intensity. The mottles are also most easily seen near the centre of the solar disk.

The mottled appearance is called the solar granulation, and the individual mottles are called

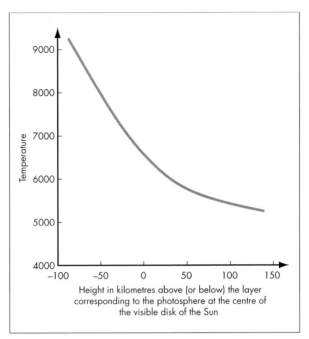

Figure 1.11. The temperature profile of the outer layers of the Sun.

granules. Granules are about 1000 km (600 miles) across, and sometimes have a polygonal shape. Their centres are hotter than the darker lanes separating the individual granules by about 400 K. The centres of granules are moving outwards and the material in the separating lanes downwards at velocities of about ± 0.5 km/s (± 0.3 miles per second). The granulation pattern is in constant motion, and individual granules form and then disintegrate over a 10–15 minute period.

The cause of the mottled appearance of the solar surface lies deep inside the Sun. The Sun produces its energy (a very tiny fraction of which is received here on Earth and keeps us going) through nuclear reactions. Those nuclear reactions occur only in the central regions of the Sun where the temperature rises to 15,000,000 K and the pressure is such that the density of the material is 14 times that of lead. The composition of the Sun at its surface is: hydrogen about 73%, helium about 25%, and all the remaining elements from lithium to uranium amounting to just 2%. The nuclear reactions at the Sun's centre convert hydrogen into helium at a rate of 600,000,000 tonnes of hydrogen to 596,000,000 tonnes of helium each second. The mass difference of 4,000,000 tonnes is converted into the 4×10^{26} J which the Sun radiates each second.[11] Although this seems a huge amount of material to be lost to the Sun, the Sun's mass at 2×10^{30} kg is such that the supply of material is sufficient to keep the Sun going for 10^{10} years. Since the Sun is currently about 4.5×10^9 years old, it will continue in much its present form for at least as long again.

Now the energy produced within the Sun's core has to travel through 700,000 km (440,000 miles) of the outer layers of the Sun before it can be radiated away. From school physics we know of only three methods whereby energy can be moved: conduction, radiation and convection.

In the Sun, and most other stars, conduction is not very important. So we are left with radiation and convection as the means of transporting solar energy towards its surface, and both have a role to play. In the central regions of the Sun, conditions do not allow convection to occur,[12] and so energy transport is just by radiation. Radiation carries the solar energy outwards from the centre of the Sun to a layer 200,000 km (120,000 miles) below the visible surface. At that point changing conditions mean that convection is no longer suppressed. From there to almost the visible surface of the Sun, most of the energy is transported outwards by convection. A few hundred kilometres below the visible surface, however, conditions change again, and so radiation predominates once more (Fig. 1.12). But the convection zone stops only just below the visible solar surface. The velocities and energy patterns within the convection zone are therefore able to penetrate through to produce effects on the visible surface of the Sun. Granules thus overlie convection cells in the convection

---

[11] The conversion is via Einstein's well-known equation: $E = Mc^2$, where E is the energy in joules, M is the mass in kg, and c is the velocity of light with the value 300,000,000 m/s. One kilogram of matter therefore converts to 10^{17} J.

[12] For convection to occur the layer must be unstable. Thus if a fragment of material is accidentally displaced upwards, say by turbulence, and is then at a higher temperature than its surroundings, it will also be less dense than its surroundings and so continue to move upwards. A portion of material displaced downwards will likewise continue to move in that direction once set in motion. For convection not to occur the temperature in the material must thus decrease only slowly with increasing height.

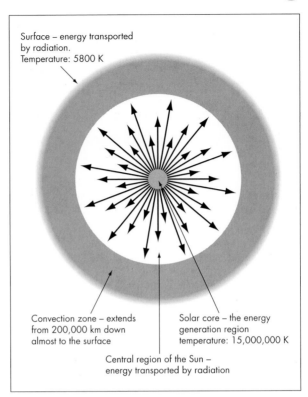

Surface – energy transported by radiation.
Temperature: 5800 K

Convection zone – extends from 200,000 km down almost to the surface

Solar core – the energy generation region temperature: 15,000,000 K

Central region of the Sun – energy transported by radiation

Figure 1.12. The internal structure of the Sun.

zone, and come and go as the convection cells rise and descend.

Faculae

There are several phenomena associated with sunspots such as plages, flares, filaments and prominences which are observable only through narrow band filters, and these are discussed in the next section and in Chapter 9. One type of feature, however, is visible in white light, and that is the faculae. These are slightly brighter patches of the photosphere which appear in and around sunspots and sunspot groups. They are at a slightly higher temperature than the normal photosphere, but only by about 300 K. They are therefore best seen close to the limb of the Sun, when their contrast is enhanced by limb darkening (Fig. 1.13). Faculae appear wherever there is a strong magnetic field, and provide a better indication of the extent of the disturbed region than do the sunspots. They usually appear before the spots and continue after the spots have disappeared.

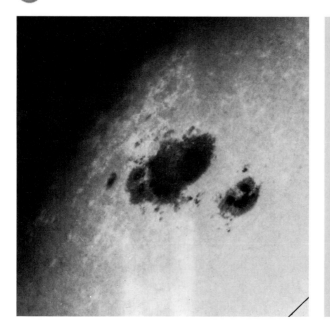

Figure 1.13. Faculae around a sunspot group near the limb of the Sun (photograph reproduced by permission of the Royal Astronomical Society).

Narrow Band Observations

Any small telescope can be used for white light solar observing just for the cost of a full aperture filter, or in some cases just by a simple adaptation to project the image (Chapter 2). There are, however, several solar features not found on white light images. Those features may be detected by observing over just a tiny fraction of the whole white light spectrum.[13] This is called narrow band observing, and involves the use of a narrow band filter, a spectrohelioscope or a prominence spectroscope (Chapter 8). These devices produce images in a single colour showing several new solar phenomena.

The new features become visible for two main reasons; the features may have been lost in the glare from the solar photosphere on white light images, whereas the single

---

[13] White light, as may be seen in a rainbow, is actually a mix of colours ranging from red through orange, yellow, green, and blue to violet. The different colours are due to differences in the wavelengths of the radiation. Red light has a wavelength of about 700 nm (30 millionths of an inch), violet a wavelength of about 380 nm (15 millionths of an inch).

colour image gives better contrast, or the single colour image may show a layer some 1000 to 3000 km above the normal solar surface with consequently different phenomena occurring from those within the photosphere.

Narrow band features can be very spectacular, and sometimes change explosively on times scales of a few minutes. There is, however, a considerable price to pay for the instruments which allow for their observation. That may be a literal price: the cost of an Hα filter (Chapter 8) is comparable with that of a 200 mm (8 inch) Schmidt–Cassegrain telescope; or it may be the time that a good DIY enthusiast would have to spend to construct an instrument such as a prominence spectroscope. None the less, narrow band features are rewarding and fascinating phenomena and the serious solar astronomer will want to observe them.

Chromospheric Network

The chromosphere can be seen at the edge of the Sun during eclipses (Chapter 7) and is a layer a few thousand kilometres thick. It derives its name from its reddish appearance and it extends upwards from a height of 500 km (300 miles) above the photosphere. It forms a reasonably continuous region up to a height of 2000 km (1400 miles) by which point its temperature has risen to 10,000 K or so. Above that it breaks up into numerous short lived, needle shaped spikes which project into the lower part of the corona. The spikes are called spicules and are typically 1000 km (600 miles) across and 5000–10,000 km (3000–6000 miles) in length. The material in them is moving upwards at a few tens of kilometres per second, and a typical spicule will last for 5–15 minutes before it fades away.

Like the photosphere, the chromosphere has a granular pattern (Fig. 1.14), but the scale of the pattern is much larger. This chromospheric network is composed of both coarse and fine mottles. The coarse mottles, which are also known as flocculi, are up to 20,000 km across, and sometimes merge to form plages. Plages are the chromospheric equivalent of faculae, described earlier. The fine mottles are a few hundred kilometres wide and can be several thousand kilometres long. They are probably spicules seen against the disk of the Sun. All the chromospheric features are associated with strong magnetic fields.

Prominences and Filaments

Prominences and filaments are different names for the same type of solar feature. They are called prominences when seen at the edge of the Sun, and filaments when

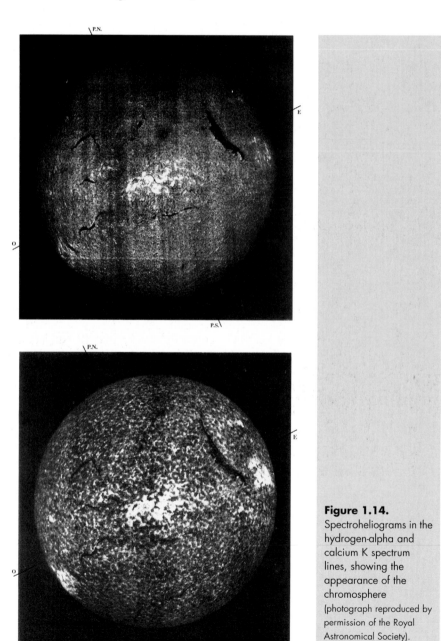

Figure 1.14.
Spectroheliograms in the hydrogen-alpha and calcium K spectrum lines, showing the appearance of the chromosphere (photograph reproduced by permission of the Royal Astronomical Society).

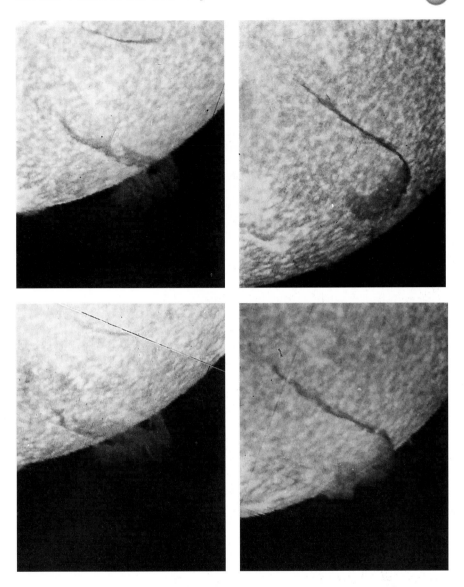

Figure 1.15. A filament/prominence (photograph reproduced by permission of the Royal Astronomical Society).

seen silhouetted as dark worm-like features against the disk of the Sun. Figure 1.15 shows the transition from filament to prominence as the rotation of the Sun carries the feature to the limb of the Sun.

Prominences are cloud-like condensations in the lowest part of the solar corona (see the section below on eclipses). They are at a temperature of about 10,000 K and have a density about a hundred times that of the surrounding corona. The temperature of the prominence is about 1% that of the corona, and so with its higher density it is in pressure equilibrium with its surroundings. The high density, however, means that the prominence is not floating in the corona in the way that clouds float in the Earth's atmosphere. For prominences to be stable, they require some form of additional support, and this is usually provided by a strong magnetic field. The shapes of prominences thus frequently mirror the shapes of magnetic fields.

Prominences range in size from a few thousand to two hundred thousand kilometres in length and are typically ten thousand kilometres high and wide, though occasionally they can extend to heights of hundreds of thousands of kilometres. Their lifetimes range from a few days to a year or more. Active prominences have rapid motions both for the prominence itself, and for the material within it. They are closely linked to complex sunspot groups. Quiescent prominences show only slow changes, such as an increase in their length. All types of prominences start life near active regions, sometimes forming before the sunspots themselves appear. The longer lived quiescent prominences often outlast the active region where they started. After the active region has disappeared such prominences slowly drift to higher latitudes, where several may merge to form a polar crown.

Flares

Flares occur within complex sunspot groups (Fig. 1.16), and are sudden rapid brightenings of a small area inside the group. They are usually only visible on single colour images, but very occasionally the biggest flares may be bright enough to be seen in white light. They also emit large amounts of energy in the microwave, ultraviolet, X-ray and gamma ray regions, and throw out very high-velocity particles which lead to solar radio bursts (Chapter 9). The total energy involved in a large flare can be up to 10^{27} J, equal to 2.5 seconds output from the whole Sun, but coming from less than 0.001% of the solar surface. Flares may last for up to an hour, but the energy is produced within the first few minutes. Temperatures as high as twenty million degrees may be reached. Condensations within the corona may follow the occurrence of a flare, and have temperatures up to four million degrees.

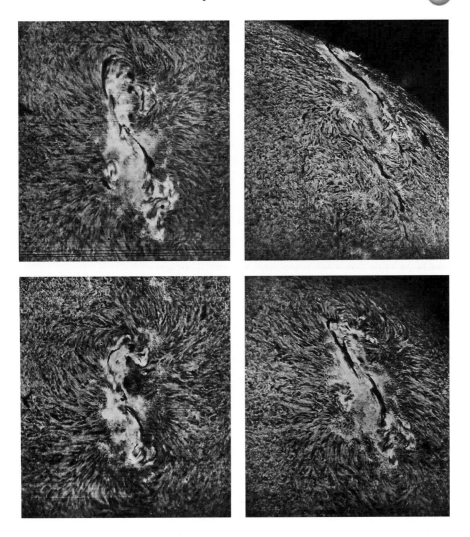

Figure 1.16. A solar flare (the bright area in the centre of the active region) seen on a calcium K spectroheliogram (photograph reproduced by permission of the Royal Astronomical Society).

Subatomic particles such as protons, electrons and helium nuclei may be ejected from flares at velocities close to that of light and lead to rapid increases in the solar cosmic ray flux at the Earth about 15 minutes after the start of the flare. The ultraviolet emission can intensify and lower the Earth's ionosphere causing short-wave radio "fade-outs". Lower-velocity particles emitted in streams may arrive at the Earth a day or so after the flare and produce strong auroral displays.

Flares are not well understood theoretically. Most hypotheses about their origin suggest that the energy involved is stored in stressed magnetic fields and released explosively as magnetic field lines reconnect, but there are still many problems in explaining all the features of flares.

Eclipses

During a total solar eclipse (Chapter 7) the chromosphere and prominences may be seen without the need for a narrow band filter (introduced in the previous section). Also the outermost layer of the Sun, the corona, becomes visible.

The corona extends outwards from the top of the chromosphere until it merges with the interstellar medium well beyond the orbit of Pluto. It has a temperature in the region of 1 million to 4 million K. The reason for this enormous temperature is still obscure. It may be the result of energy being transported out from the convection layer of the Sun (Fig. 1.12) by shock waves or magnetic fields, or it may be linked with the solar rotation. The density of the corona, however, is so low[14] that despite its extremely high temperature, it is faint and can only be seen during total eclipses or through the use of special instruments such as coronagraphs (Chapter 8).

The shape of the corona varies with the sunspot cycle. It often develops long streamers above active regions which extend to two or three times its normal limit (Fig. 7.7). The high temperature of the corona means that it is too hot to be held by the solar gravitational field. The material in the corona is thus continually expanding out into space and being lost to the Sun and solar system. The expanding corona is observed near the Earth as the solar wind, and particles are streaming past the Earth at velocities of around 500 km/s (300 miles per second).

Spectra of the corona show that it is formed from three components: the K, F and E coronae. The K corona has a uniform spectrum. It is the result of light from the solar photosphere being scattered by electrons within the corona. At coronal temperatures, the electrons are moving at velocities around 10,000 km/s (6000 miles per second). The absorption lines[15] in the

---

[14] 0.0000000000001% of the density of water at a height of one solar radius above the photosphere.

[15] When the spectrum of the Sun (or star, galaxy, planet, etc.) is observed, it is found that the bright rainbow colours have numerous narrow dark regions crossing them. These are called spectrum absorption lines. Every element (hydrogen, helium, oxygen, iron, etc.) when in gaseous form will produce a distinctive pattern of absorption lines which is unique to that element.

scattered solar radiation are therefore smeared out by the Doppler shifts[16] induced by the moving electrons, and only a featureless continuum spectrum is left to be seen.

The F corona has a spectrum almost identical to that of the Sun. It is solar light scattered by dust particles. The dust particles move slowly compared with the electrons, and so there is no Doppler smearing to eliminate the absorption lines. The dust is concentrated in the plane of the solar system, and the same effect can be observed at night as the zodiacal light. This latter is a faint glow which can be seen soon after sunset or before sunrise on clear moonless nights from a good observing site. The glow runs along the ecliptic[17] and rises up from the horizon from near the point at which the Sun set, or from which it will rise.

The E corona is the result of emission by the material of the corona itself, not from the scattering of solar light. It has an emission line spectrum[18] which arises from the atoms and ions forming the corona.

The corona as a whole doubles in intensity between sunspot minima and maxima, and most of this change is due to the intensification of the K corona. At sunspot minima, the K corona is about three times brighter than the F corona and it is about six times brighter at sunspot maxima. The E corona has an overall intensity

---

[16] If the relative line-of-sight velocity between an object and the Earth is towards each other, then the observed wavelengths of light from the object become smaller than those measured in the laboratory. If the movement is apart, then the wavelengths become longer. This is the Doppler shift. The wavelength changes enable the velocity between the Earth and the object to be determined through the use of the Doppler formula;

$$\text{Velocity} = 300,000 \frac{\lambda_{\text{Observed}} - \lambda_{\text{Laboratory}}}{\lambda_{\text{Laboratory}}} \quad \text{km/s}$$

where $\lambda_{\text{Observed}}$ is the observed wavelength and $\lambda_{\text{Laboratory}}$ is the corresponding laboratory wavelength.

[17] The ecliptic is the path of the Sun across the sky throughout the year. It runs through the centre of the zodiac, and the movements of most of the planets closely follow the ecliptic.

[18] An emission line spectrum is the inverse of an absorption line spectrum. Elements in gaseous form emit light at the same wavelengths that they absorb it, producing bright lines in the spectrum when the gas is silhouetted against a dark background.

only about 1% that of the combined K and F coronae, but since it is composed of emission lines it can be proportionally much brighter than this at the specific wavelengths corresponding to those lines.

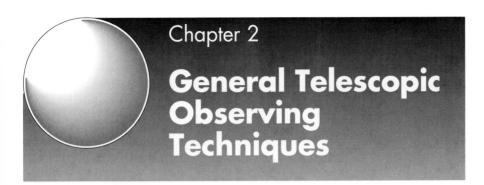

Chapter 2

General Telescopic Observing Techniques

CAUTION REQUIRED

Note the warning at the start of this book, and do not attempt any solar observation until you have assimilated all the details in this and other chapters.

To repeat: smoked glass and exposed film **cannot** be relied upon to act as safe filters for looking directly at the Sun; neither can you depend on CDs, space blankets, aluminised helium balloons, potato crisp packets, floppy disks, smoked plastic, crossed Polaroid filters, sunglasses, mirrors, medical X-rays or almost any other of the items that may be suggested from time to time as a substitute for the filters discussed below, to act as a reliable filter.

Remember that you may have more than one telescope. Most main telescopes have a small telescope attached to their side to act as a finder, and in some cases there may be more than one. The objective of the finder must be blanked off (covered completely with an opaque screen), or equipped with a full-aperture filter **before** any solar observing procedure is attempted with the main telescope. The sunlight passing through a finder if it is not blanked off may be sufficient to cause burns or to damage parts of the telescope. Note also that most finders have cross-wire eyepieces to enable them to be used for aligning the main telescope onto an object. The focused image of the Sun is likely to melt or burn these cross-wires, so the finder should not be used for eyepiece projection unless a non-cross-wire eyepiece is put in place of the normal finder eyepiece (see the section on projection below for further details).

Full Aperture Filters

There are two main ways in which it is safe to observe the Sun through a telescope or binoculars. One is through the use of a suitable solar filter, the other by projecting the image. The first of these is the concern of this section, the second is considered in the next section.

If you are leading a group of non-astronomers or a class of children in observing the Sun, then it is generally better to view the projected solar image. This is so that no one looks anywhere near the Sun in the sky. Children especially may be tempted to go off and try things out for themselves with inappropriate filters or by squinting through almost closed fingers, etc., with possible disastrous consequences. If you are leading, or in any other way involved with, such groups then it is very strongly recommended that **only** projection methods are used. However note that not all telescopes can be used for projecting the solar image (see next section). Designs such as Schmidt–Cassegrains, which have a short focal ratio primary mirror, can produce internal images of the Sun of sufficient intensity that they may cause damage to the telescope. Always check the manufacturer's instructions, or contact them directly, to check your telescope's suitability before using it to project the solar image. With some telescope designs and manufacturers, the guarantee may be invalidated by using the telescope for solar work without a full aperture filter in place.

CAUTION
REQUIRED

Filter Basics

If you do choose to look at the Sun through a filter, then you must ensure that the filter eliminates the ultraviolet and infrared components of the solar radiation as well as reducing the visible light intensity. There are no American or international standards for solar filters, but the European Community directive on personal protective equipment (EN172) recommends the following minimum requirements for unaided viewing:

Wavelength	Transmission (%): see Table 2.1
280–380 nm (ultraviolet)	• 0.003%
380–780 nm (visible)	• 0.0032%
780–1400 nm (near infrared)	• 0.027%

The directive also requires filters to be large enough to cover both eyes simultaneously, and so have a minimum size of 35 mm × 115 mm (1.4 in × 4.6 in).

Filters are often characterised by their optical density, which is denoted by *Den*. This expresses the reduction in the intensity of a beam of radiation passing through the filter as a power of 10, with the power involved being the optical density. Thus an optical density of 1 corresponds to a reduction in intensity by a factor of ×10 (= 10^1), an optical density of 2 corresponds to a reduction in intensity by a factor of ×100 (= 10^2), an optical density of 3 corresponds to a reduction in intensity by a factor of ×1000 (= 10^3), and so on (Table 2.1).

Shade numbers are also used to characterise filters. The shade number is given by:

$$\text{Shade number} = (7/3)\log_{10}(1/T) = (7/3)\log_{10}(100//T\%) = (7/3)\,Den$$

where T is the actual visible light transmission and $T\%$ is the percentage visible light transmission.

Table 2.1. Optical Density

Optical Density (Den)	Reduction in intensity	Transmitted radiation (% of incident radiation)	Shade number
0.5	×3	33%	1.2
1.0	×10	10%	2.3
1.5	×30	3%	3.5
2.0	×100	1%	4.7
2.5	×300	0.3%	5.8
3.0	×1,000	0.1%	7.0
3.5	×3,000	0.03%	8.2
4.0	×10,000	0.01%	9.3
4.5	×30,000	0.003%	10.5
5.0	×100,000	0.001%	11.7
6.0	×1,000,000	0.0001%	14.0
7.0	×10,000,000	0.00001%	16.3
8.0	×100,000,000	0.000001%	18.7
2.6	×400	0.27%	6.0
3.4	×2,500	0.037%	8.0
4.3	×20,000	0.0052%	10.0
5.1	×130,000	0.00072%	12.0
6.0	×1,000,000	0.0001%	14.0
6.9	×8,000,000	0.000014%	16.0

Thus a shade number of 12 corresponds to a percentage transmission of 0.00072%. Shade numbers are also listed in Table 2.1. Two filters used in combination will have a shade number equal to the sum of their individual shade numbers. Thus if #6 and #10 filters are used together, they would be equivalent to a single #16 filter.

From the table therefore we may see that a solar filter needs a density of at least 4.5 in the ultraviolet and visual regions and 3.5 in the near infrared. In practice the only certain way of accomplishing this is to use a professionally manufactured filter. There are two suitable types; custom-designed filters sold by reputable astronomical or nautical equipment suppliers specifically for solar work, and welder's goggles. Shade numbers of welder's filters of 12 and higher are safe (for further details see Chapter 6).

Telescope Basics

It might seem from the above discussion on the use of filters for unaided observation of the Sun that filters of quite enormously high densities would be needed when using a telescope. However that is not the case – the filters suitable for naked eye observations are also suitable as full aperture filters on telescopes. The reason why denser filters are not needed is that a telescope does not increase the surface brightness of an extended object[19] (like the Sun) over that seen by the unaided eye. More energy is of course collected by the larger aperture of the telescope compared with that of the eye, but the increased magnification produced by the telescope then spreads that energy out over a larger area. A precise calculation[20] shows that the resulting

---

[19] Point sources, like stars, are, however, brighter when seen through a telescope.

[20] Interested readers may find the complete calculation in Chapter 2 of *Telescopes and Techniques* by C. Kitchin, published by Springer-Verlag, 1995. Three of the author's books may provide useful reference material for this book (see also Appendix 1): they are the above book (abbreviated to *T&T*), *Astrophysical Techniques*, 3rd edition, published by the Institute of Physics, 1998 (abbreviated to *AT*), and *Optical Astronomical Spectroscopy*, published by the Institute of Physics, 1995 (abbreviated to *OAS*).

Table 2.2. Relative Surface Brightness of the Solar Image
NB The low percentage values towards the bottom of this table **do not** make it safe to look through the unfiltered telescope: see European safety directive, EN172 (described in the previous section). Nor should the term "brightness to the unaided eye" be taken to imply that it is safe to look at the Sun with the unprotected naked eye: see Chapter 6.

Magnification (D is the diameter telescope in metres)	Magnification (D is the diameter of the telescope in inches)	Surface brightness of the Sun through the telescope compared with the surface of the brightness to the unaided eye (%)
×50D	×1.25D	100
×75D	×1.88D	100
×100D	×2.5D	100
×125D	×3.13D	100
×140D	×3.5D	100
×150D	×3.75D	89
×175D	×4.38D	65
×200D	×5.00D	50
×250D	×6.25D	32
×300D	×7.50D	22
×350D	×8.75D	16
×400D	×10.0D	13
×500D	×12.5D	8
×600D	×15.0D	6
×750D	×18.8D	4
×1000D	×25.0D	2
×2000D	×50.0D	0.5
×4000D	×100.0D	0.1
×6000D	×150.0D	0.06
×8000D	×200.0DD	0.03
×10,000D	×250.0DD	0.02

image at its brightest has a surface brightness equal to that seen by the unaided eye. Most of the time the telescopic image will actually have less energy per unit area than that seen directly (Table 2.2).

Thus from the table we may see that the surface brightness of the Sun is reduced to 0.5% of its naked eye value when a 50 mm (0.05 m – 2 inches) diameter telescope is used at a magnification of ×100.

This consideration of surface brightness is relevant to the potential for damage arising from the focused solar image, i.e. potential damage to the retina of the eye. A filter that is safe for naked eye viewing of the Sun when used as a full aperture filter (see below) on a telescope will also reduce the intensity of the image on the retina to safe levels. **However,** the telescope, even with such a filter in place and whatever the magnification, still increases the total amount of energy entering the eye. The maximum safe diameter even with an appropriate

CAUTION
REQUIRED

full aperture filter on the telescope is therefore about 150 mm (6 inches). Larger instruments should always be stopped down to 150 mm (6 inches) or to smaller sizes.

Full Aperture Solar Filters

Full aperture filters, as their name implies, are filters which cover the whole aperture of the telescope (Fig. 2.1). The light and heat intensity from the Sun is therefore reduced **before** it is concentrated by the telescope optics. Filters should **never** be used at the eyepiece end of the telescope, since the concentration of energy may be sufficient to melt or crack them, with disastrous consequences if you are looking through the telescope at the time. Also the internal image formed by the objective may cause damage to the structure of the telescope (see the next section).

The term "full aperture filter" as used here also includes filters at the entrance aperture of the telescope which do not cover the whole aperture. Since large telescopes should be stopped down (see above) sub-aperture filters are quite widely encountered. When using such subaperture filters, the uncovered areas of the telescope aperture must be blanked off with a completely opaque screen. The filters are therefore normally mounted onto a suitable sheet of metal or plywood with a hole into which the filter fits. The mounting sheet and filter together then cover the whole of the telescope entrance aperture. Used in this manner, subaperture filters behave essentially as full aperture filters.

Figure 2.1. The full aperture filter: left, mounted ready for attachment to a telescope; right, sheet film for mounting into an adaptor for the telescope.

One of the most widely available full aperture filters is formed from a thin Mylar sheet, overcoated with an aluminium, chromium, stainless steel or other metallic film. For most purposes, and certainly all visual work, the double layer design is best. This uses two over-coated plastic films with the metal coatings sandwiched in the centre. This not only provides protection for the comparatively fragile metal coatings, but any pin-hole left by the coating process on one side is almost certain to be covered by an intact area of the other coating. A recently introduced version of these filters produced by Baader claims to produce better quality images than the Mylar-based filters, but details of its construction are not provided. It is, however, available in loose sheet form at low cost (£10–15, $15–25, for an A4 sized sheet) for DIY construction of a full aperture filter. Versions of the filters with a single metal coating may be found and these are more suited to solar photography (Chapter 4) since they have a higher transmission than the double layer versions. The filters are sold under various trade names, and an inspection of the advertisements in any popular astronomy magazine will soon supply the contacts for several suppliers.[21] A list of some solar filter suppliers current at the time of writing is given in Appendix 2. The optical density of the filters varies with the supplier, but typically has values of 4.5–6 over the ultraviolet and visible parts of the spectrum and 2.5–4.5 in the near infrared. The transmission curve of a typical filter of this type is shown in Fig. 2.2.

The filters often have quite a wrinkled appearance (Fig. 2.3) and to observers accustomed to having optical surfaces accurate to a small fraction of the wavelength of light, they seem unlikely to give good images. The plastic, however, is sufficiently thin (its thickness is around 5 microns, or 0.005 mm or 0.0002 inches), that very little distortion of the solar image occurs in practice. One supplier guarantees a resolution for the telescope with the filter in place which is at least 90% that of the unfiltered telescope. The filters are quite fragile and need to be handled carefully. Pin-holes can be repaired with opaque cloth-based sticky tape, but if a filter becomes more damaged than this, then it must be discarded, and a new one purchased.

---

[21] Details may also be found via the internet: search for "solar filter", or check the "Sky and Telescope" resources page.

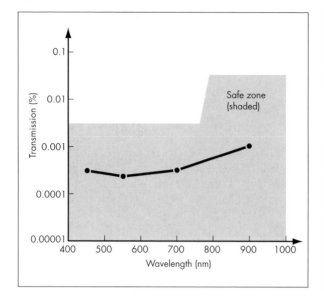

Figure 2.2.
Transmission curve for a metal-on-plastic solar filter.

Figure 2.3. A 125 mm (5 inch) metal-on-plastic solar filter on a 180 mm (7 inch) Maksutov telescope. Note that the area of the telescope not covered by the filter has been blanked off completely.
NB In this instance (and also in several later images) the filter has been mounted onto the telescope using thick stiff black foam. This is convenient since it may easily be shaped to fit the filter and the telescope. However the foam **must** be opaque, and **must** be securely attached to the instrument. Thin, flexible, clear or white foam, or any type of foam through which even a small fraction of light passes **must not** be used in this fashion. Other suitable materials to mount filters and to blank off parts of the telescope objective include plywood, sheet metal and thick cardboard.

A related and increasingly popular filter has a metal coating on a glass substrate. The substrate needs to be of optical quality if the image is not to be distorted. These filters therefore tend to be more expensive than the metal-on-plastic variety, although the prices are becoming more comparable. Typically though the prices of metal-on-glass are 50–100% more than metal-on-plastic for a given size. They are, however, more rugged than the metal-on-plastic filters previously discussed. The metal coating can be aluminium, stainless steel or a nickel–chrome–iron alloy known as Inconel. They are used in a similar manner to the metal-on-Mylar filters. Some versions have a slightly higher transmission than others, and so are more suited to photography than for visual work. These are identified by some suppliers who class their filters as "visual" or "photographic". Metal-on-glass filters do not introduce polarisation into the solar image, which can be an advantage for autofocusing cameras (see below). Internal reflections, however, may occur between the two surfaces of the filter leading to faint ghost images near the main image (it will make no difference which way the filter is placed on the telescope, since the coating will reduce the intensities of both the direct image and the ghosts by similar factors). Although the reasons are not entirely clear, metal-on-glass filters do tend to show sunspot detail better than metal-on-plastic filters, while the latter are often better for faculae.

The solar image seen through either metal-on-Mylar or metal-on-glass filters is usually coloured. Blue or orange-yellow are the normal shades produced, although fawn and brown may also be encountered. Inconel-based filters and the Baader product previously mentioned are close to neutral in their effects, but the latter still has a slightly pinkish tinge. All other things being equal, better images will be obtained with filters that give a yellow or red image, than those which give blue images, since the scattering in the Earth's atmosphere increases as the wavelength decreases. This effect, however, is usually swamped by other problems, and so unless you are making photometric measurements at different wavelengths, the colour cast of the image is largely only of aesthetic significance. Many people prefer the orange shade, regarding the blue images as having a "cold" appearance, but it is a matter of personal preference. For monochromatic imaging the colour produced by the filters is of no significance. For colour photography, however, a more realistic appearance (i.e. a white image) may be realised by combining the solar filter with an appropriate colour filter. A blue transmission filter will help to correct the yellow-orange images and orange-red filter the blue images. The colour on CCD images can be corrected by image processing after the event (Chapter 4).

For photometric work, the use of these filters will cause problems because the colour of the solar filter will upset the normal balance between the photometric filters. The transmission of the solar filter at the wavelengths of the photometric filters in use will need to be known accurately, so that its effects may be corrected. Transmission curves for the solar filters may be available from their manufacturers. Alternatively the corrections may need to be determined experimentally by measuring the outputs from an artificial light source first through just the photometric filters, and then through the photometric filters and the solar filter together.

The plastic support of the metal-on-Mylar filters may induce some polarisation into the solar image. This will not normally lead to problems unless you are using polarising filters elsewhere within the system or are using a detector, such as a photomultiplier, which may vary its sensitivity depending on the direction of the polarisation. The detection of magnetic fields on the Sun depends upon measuring polarised light (Chapter 10), but the techniques involved with such measurements are unlikely to be used by amateur astronomers. Polarisation of the light may also affect the focusing of some autofocus cameras, and so cause problems if these are used for magnified imaging (Chapter 4). The induced polarisation is a result of the manufacturing processes involved in producing plastic sheet; it is not due to the coating. Metal-on-glass filters are therefore free of polarisation effects and may be preferred for use with cameras.

In use, the full aperture filter is positioned so that the whole entrance aperture of the telescope is completely covered. If your filter is smaller than your telescope, then the exposed areas of the aperture **must** be covered with an opaque screen (Fig. 2.3). The filter and any additional screen must be securely attached to the telescope, so that there is no possibility of it falling off or being dislodged by the wind. Only when the filter is securely in place is the telescope then pointed towards the Sun. **NB** When using a full aperture filter do not forget to blank off your finder telescope, or its projected solar image can produce a nasty burn if it falls onto your skin.

Initial alignment on the Sun is best accomplished by circularising the shadow of the telescope (Fig. 2.6). In contrast to the eyepiece projection method (see next section), however, the telescope is safe to look through when the filter is in place, in order to centralise the image. If you have a full aperture filter for the finder telescope as well as for the main telescope, then the instrument can be pointed roughly in the right direction, using shadows, and the Sun then found with the finder telescope as for normal night-time observation. However, if using the finder in this way, or a small diameter telescope of any type, then it should be provided with a cardboard screen so that the eye not being used to look through the telescope cannot accidentally be exposed to the Sun.

Observation of the Sun can proceed like a normal night-time observation when using a full aperture filter. You will probably, however, find that it is best to focus on the edge of the Sun at the start of the observing session. Sunspots and other features are small compared with the Sun itself, and (unlike, say, lunar craters) are not to be found over the whole Sun. Once the image is sharply focused, however, the spots, plages, faculae, etc. (Chapters 1 and 5) can be found by scanning the main disk of the Sun. Any available eyepiece can be used, but very low magnifications should be avoided so that the solar surface brightness is reduced (Table 2.2). Very high magnifications are also unlikely to be usable because turbulence in the Earth's atmosphere caused by the heat from the Sun will make for very poor images. Photographs and CCD images can also be obtained in the usual way if so wished (Chapter 4).

Binoculars can be used for observing the Sun if they are fitted with full aperture filters, though at the commonly available magnifications ($\times 7$ and $\times 10$) relatively little detail will be seen. They are, though, ideal for following the progress of the Moon over the Sun during eclipses (Chapter 7). To use binoculars for solar observing, either both main lenses (objectives) of the binoculars must have separate full aperture filters, or a single filter must cover both lenses, or the side of the binoculars without the full aperture filter must be completely covered by an opaque screen. Welder's number 14 filters (Chapter 6), though of variable optical quality, may be good enough to be used as full aperture filters on binoculars (Fig. 2.4). The filter, however, must be securely mounted on the binoculars and of shade number 14 or higher.

Figure 2.4. A welder's number 14 filter mounted on a pair of binoculars. Note that the unused half of the binocular has been blanked off with an opaque screen. **NB** See the warning about the use of foam for mounting filters with Fig. 2.3.

Projection

Warning: the alternative safe observing technique to the use of full aperture filters is to project the solar image. As already noted, when demonstrating to a group of people, especially if the group includes children, projection methods are better since they do not require looking anywhere near the Sun in the sky. **However**, projection methods do carry their own hazards. In particular a focused image of the Sun is usually formed inside the telescope, and this can damage the instrument (see later discussions in this section). This is a significant problem with telescope designs such as the Cassegrain, Schmidt–Cassegrain and Maksutov which have short focal ratio[22] primary

CAUTION
REQUIRED

---

[22] The focal ratio, also known as the f-ratio, is the focal length of the lens or mirror divided by its diameter. A refractor (lens-based telescope) will normally have an f-ratio of f10 to f15, a Newtonian telescope one of between f8 and f10. The final f-ratio for a Schmidt-Cassegrain or Maksutov telescope is likely to be around f10, but the primary mirror may be f3 or f4. Camera lenses sometimes have f-ratios down to f1.8 or even f1.5.

mirrors. Any design of telescope may have internal stops and/or shade tubes to cut down on background light. Since the Sun is 0.5° across, parts of its image may overlap onto and damage such structures even when the image is centred within the telescope. If stops or shade tubes are a part of your telescope, or you have any other doubts about using it for solar work, then check on its suitability with the manufacturer before using the telescope to project the solar image. With some telescope designs the manufacturer's guarantee may be invalidated if the telescope is used to project the solar image.

Remember also that your finder telescope may have a cross-wire eyepiece, and those cross-wires may be melted or burnt by a focused solar image. Do not therefore use your finder for eyepiece projection without replacing its cross-wire eyepiece, and under all other circumstances make sure that the finder is completely blanked off before undertaking any solar work.

Whatever your telescope, **always** read and follow the manufacturer's instructions as well as the guidance given here before starting observations.

Telescopes

In contrast to most other areas of astronomy, there is no problem with lack of light when observing the Sun. For most purposes a 75–100 mm (3–4 inches) or smaller aperture will be sufficient. Owners of larger telescopes will therefore need to reduce the effective size in order to undertake solar work. This is most simply done by stopping the telescope down.

Stopping down just involves placing an opaque screen of cardboard, thin plywood or metal, which has a hole 75–100 mm across cut in it, over the telescope's objective (Fig. 2.5). Only the light passing through the hole is then received by the telescope. For reflectors with a secondary mirror, the hole in the screen will need to be placed off-axis so that it is not obscured by the secondary mirror. Care should be taken to ensure that the screen is firmly attached for if it were to fall off or blow away, the full aperture of the telescope would suddenly be gathering sunlight with possibly disastrous results to the observer or instrument.

Figure 2.5. A 180 mm (7 inch) Maksutov telescope stopped down to 100 mm (4 inches) aperture for eyepiece projection (effective aperture is 80 mm because of the secondary mirror obstruction). **NB** See the warning about the use of foam for mounting filters with Fig. 2.3.

First stop your telescope down to an aperture of 75–100 mm (3–4 inches) or so. Then securely cover the finder telescope if there is one attached to the main telescope. Put an inexpensive, low power eyepiece into the main telescope. Such an eyepiece is needed because even stopped down, a lot of energy will be gathered by the telescope. This could heat the eyepiece to the point where it might be damaged. It is therefore wise not to use an expensive eyepiece for eyepiece projection. Furthermore the more expensive eyepieces are likely to have more lenses, and to be more complex in construction than simpler eyepieces; they are therefore more likely to be vulnerable to heat damage. **NB** As an additional point, do not use an eyepiece which has cross-wires within it for guiding; the energy from the Sun may well distort them or even melt or burn them.

CAUTION
REQUIRED

Next point the telescope at the Sun – **but you must do this without looking through the telescope.** The simplest method of finding the Sun without looking through the telescope is to circularise the shadow of the

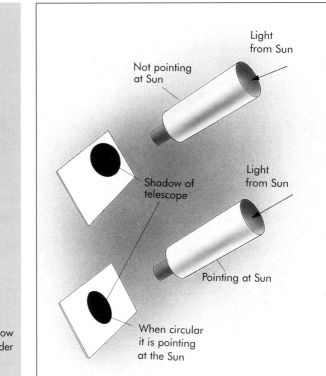

Figure 2.6.
Circularising the shadow of the telescope in order to find the Sun for projection (see also Fig. 6.4).

telescope (Fig. 2.6). First point the telescope roughly towards the Sun, and then look at its shadow on the wall, the ground, or on a suitable board placed behind the telescope. Then move the telescope, without looking through it, so that the shadow decreases in size. When the telescope is pointing at the Sun, the shadow will be at its minimum size, and square-on to the telescope will be circular (or whatever may be the shape of the cross-section of the telescope).

Probably while setting the telescope on the Sun through circularising its shadow, you will have had a brief glimpse of the solar image zipping through the telescope's shadow. With luck, when you have finished setting the telescope, some of the solar image will be visible within the shadow as part of a brighter circle. If no part of the solar image can be seen, then the telescope should be moved by small amounts, until it does appear. **Do not** be tempted to look through the telescope if you have difficulty locating the solar image

– recircularise the telescope's shadow and persist with nudging the telescope. The image will appear eventually. Once a part of the image is visible, then it may be centred on the projection screen using the telescope's slow motions (Fig. 2.8). The solar image is then ready for projection.

Owners of Cassegrain, Schmidt–Cassegrain and Maksutov telescopes, or any other design in which the primary mirror has a small focal ratio, need to take extra care whilst finding the Sun (**NB** See also the warning above about invalidation of some manufacturer's guarantees). The primary mirror of such telescopes can form an internal image which has sufficiently high intensity to cause damage to the structure of the telescope, or even to cause it to catch fire, if that image falls onto parts of the telescope that can absorb heat, during the process of finding the Sun. The telescope should therefore be stopped down to no more than 50 mm (2 inches), and preferably to 25–30 mm (1–1.2 inches), while the image of the Sun is acquired. Once the image has been centred, the larger stop can be substituted for the smaller one. However, take care not to allow the full aperture of the telescope to be exposed to the Sun while swapping the stops. A smaller stop which may be inserted inside the larger one (Fig. 2.7) is thus probably better than two separate stops.

CAUTION
REQUIRED

Care needs to continue to be taken when using the larger stop to ensure that the solar image remains centred. If the telescope is not driven at the solar rate[23] or does not have a drive at all, then the image will drift and can soon overlap onto and damage the internal structure of the telescope. This can be a worse problem than that which occurs during finding the Sun, since the image may drift slowly and so remain on one part of the telescope for some time.

---

[23] The Sun moves from day to day around and also up and down in the sky. The normal drive for a telescope therefore does not allow the telescope to track the Sun correctly. The Sun's movement around the sky ranges from 0.896°/day to 1.113°/day from west to east. The telescope drive must therefore be adjusted to be between 0.25% and 0.31% slower than normal if it is to track the Sun. The Sun's vertical movement, however, is not corrected by most telescope drives and this can be up to 0.395°/day.

Figure 2.7. A 25 mm (1 inch) finding stop in place on a 180 mm (7 inch) Maksutov telescope. **NB** See the warning about the use of foam for mounting filters with Fig. 2.3.

Any design of telescope may have internal stops and/or shade tubes which are intended to cut down background light during normal observing. Since the Sun is 0.5° across, parts of its image may fall onto these stops even when that image is centred in the telescope (see the warning above). If your telescope includes stops or shade tubes, check with its manufacturer before using it to project a solar image. Even if the telescope is not damaged, its guarantee may be invalidated if it is used against the manufacturer's guidelines.

Having found the solar image as described above, hold a sheet of white cardboard 0.2–0.5 m (8–20 in) behind the eyepiece. You should then see at least a part of the projected image of the Sun (Fig. 2.8). Move the telescope until the whole disk of the Sun can be seen, and adjust the eyepiece (while looking at the screen – **do not** look through the eyepiece) until the image is focused. Features such as sunspots, limb darkening and possibly faculae and granulation (Chapters 1 and 3) should then easily be visible on the projected image.

If you intend observing the Sun frequently, then it is worth constructing a framework to attach to the telescope to hold the screen onto which the image is projected (Figs. 2.8 and 2.9). A simple cardboard shield to prevent direct solar light from falling onto the screen will also help to improve the image by increasing its contrast.

The image of the Sun obtained by eyepiece projection can easily be sketched. The outlines of the solar disk and its features can be traced directly from the image onto a piece of paper placed onto the projection screen.[24] Fine detail can then be added when the paper is removed from the screen. The image is also usually easily bright enough to be photographed (see also Chapter 4). If the camera, however, is held to one side of the screen, then the image will be slightly distorted by foreshortening. With care, however, the screen can be tilted slightly without the image going out of focus. If the camera is held at an equal angle on the opposite side of the screen to the telescope, then an undistorted image will be obtained (Fig. 2.9). With electronic images, simple image processing (Chapter 4) will restore a circular image.

The size of the image of the Sun obtained by eyepiece projection depends upon the focal length of the telescope, that of the eyepiece and the projection

Figure 2.8. Eyepiece projection of the solar image.

---

[24] This is the same principle as the way in which the camera obscura was used in the eighteenth and nineteenth centuries to sketch the outlines for landscape paintings.

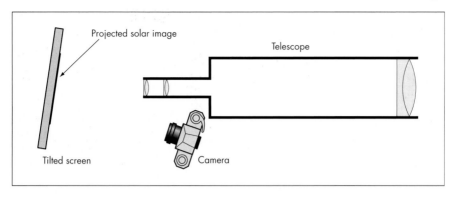

Projected solar image

Telescope

Tilted screen

Camera

Figure 2.9.
Obtaining undistorted photographs of a solar image produced by eyepiece projection.

distance (the distance from the eyepiece to the screen). The focal length of the telescope will usually be marked on the instrument. Sometimes, however, the diameter and focal ratio will be given instead. The focal length is then found by multiplying these two figures together. Thus a 200 mm f10 telescope has a focal length of 2000 mm (2 m or 80 inches). Table 2.3 gives the size of the solar image for various combinations of telescope and eyepiece focal lengths and projection distance. The image size for other combinations may be found by interpolation. Thus for the above telescope with a 25 mm (1 inch) eyepiece and a 0.5 m (20 inch) projection distance, the solar image would be 372 mm (15 inches) across. Using the same eyepiece with a projection distance of 0.4 m (16 inches) would give an image 298 mm[25] (12 inches) across, and so on.

Binoculars

CAUTION
REQUIRED

It is recommended that binoculars are **not** used for solar projection, since the heat involved may cause damage to the cement used to hold the internal prisms in place, or to other parts of the instrument. If you do decide to use binoculars for eyepiece projection then check on their suitability with their manufacturer before doing so. The procedure for eyepiece projection using binoculars is the same as that for telescopes

---

[25] $= 372 \times 0.4/0.5$.

Table 2.3. The size (in mm) of the projected solar image

Size of the projected solar image (mm)		Telescope focal length (m)									
Eyepiece focal length (mm)	Projection distance (m)	0.25	0.5	0.75	1	1.5	2	2.5	3	5	10
40	0.25	15	29	44	58	87	116	145	174	291	581
40	0.50	29	58	87	116	174	232	291	349	581	1162
40	0.75	44	87	131	174	262	349	436	523	872	1744
40	1.00	58	116	174	232	349	465	581	698	1162	2325
25	0.25	23	46	70	93	140	186	232	279	465	930
25	0.50	46	93	140	186	279	372	465	558	930	1860
25	0.75	70	140	209	279	418	558	698	837	1395	2790
25	1.00	93	186	279	372	558	744	930	1116	1860	3720
12.5	0.25	46	93	140	186	279	372	465	558	930	1860
12.5	0.50	93	186	279	372	558	744	930	1116	1860	3720
12.5	0.75	140	279	418	558	837	1116	1395	1674	2790	5580
12.5	1.00	186	372	558	744	1116	1488	1860	2232	3720	7440

except that the aperture should be stopped down to no more than 20 mm (0.8 inches), and the main lens (objective) of the half of the binoculars not in use **must** be covered completely with an opaque screen.

Other Approaches

Reflection

Glass, plastic, water and other so-called transparent materials actually reflect a proportion of the light that falls onto them. The amount of the light that is reflected depends upon the angle at which it falls onto the surface.[26] If the angle is very small, when it

CAUTION REQUIRED

---

[26] The reflectivity is given by Fresnel's equation:

$$\frac{R}{I} = 0.5\left[\left(\frac{\sin(i-t)}{\sin(i+t)}\right)^2 + \left(\frac{\tan(i-t)}{\tan(i+t)}\right)^2\right]$$

where R is the reflected intensity, I is the incident intensity, i is the angle of incidence measured to the perpendicular to the

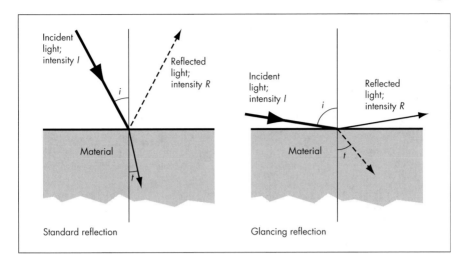

Standard reflection Glancing reflection

Figure 2.10.
Reflection.

is known as glancing incidence (Fig. 2.10), any substance becomes a good reflector. Even matt black cardboard will reflect at low enough angles, as you may easily try for yourself. At higher angles of incidence, however, the reflectivity of transparent materials falls to a small percentage (Table 2.4). Reflection from plain glass or from a still water surface may therefore be used to reduce the intensity of light from the Sun. Both the visible and the infrared intensities are reduced by reflection. A single reflection, however, is **not** sufficient for safe observing, and it needs to be combined with a solar filter. Low angle reflections, such as might come from looking at water in a distant lake or pond, must be completely avoided, since the reflectivity then rises sharply, approaching 100% at very small angles (Table 2.4).

(*footnote 26 continued*)
surface (Figure 1.13), and t is the angle that the transmitted light makes to the perpendicular. When *i* is near zero the formula simplifies to

$$\frac{R}{I} = \left[\frac{n-1}{n+1}\right]^{-2}$$

where *n* is the refractive index of the material.

Table 2.4. Reflectivity of uncoated materials (i is the angle of incidence of the solar light onto the material, i.e. the angle to the perpendicular to the surface)

Material	$i = 0°$ (%)	$i = 45°$ (%)	$i = 85°$ (%)
Crown glass	4	5	61
Flint glass	6	7.5	62
Polycarbonate plastic	4	5	61
Water	1.7	2.5	57

Solar Diagonals

In the past, the reduction in intensity caused by reflection from plain glass has been used as the basis of a device for observing the Sun known as a solar diagonal or Herschel wedge (Fig. 2.11). These devices are **not** now recommended for observing the Sun. The reason for the change in the recommendation on the use of solar diagonals is that in order to be safe, they must be used on a 50 mm (2 inch) or smaller telescope, and at a minimum magnification of $\times 300$. This arises because plain glass reflects about 5% of the incident radiation, and the solar intensity must therefore be reduced by a further factor of $\times 1700$ (a comparative brightness reduction to 0.06%) in order for the final intensity of the image to reach the safe limit of 0.003% (see full aperture filters). From Table 2.2 this gives the minimum required magnification of $\times 6000D$, (i.e. $\bullet$ $\times 300$ for the 50 mm telescope). Since nowadays few

CAUTION
REQUIRED

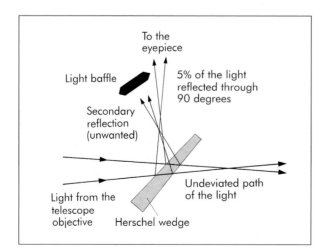

To the eyepiece

Light baffle

5% of the light reflected through 90 degrees

Secondary reflection (unwanted)

Light from the telescope objective

Herschel wedge

Undeviated path of the light

Figure 2.11. The solar diagonal or Herschel wedge.

people use telescopes as small as 50 mm except as finders, and rarely will seeing condition in the daytime be good enough to allow such high magnifications, there is thus a great temptation to use the solar diagonal on a telescope that is too large and with too low a magnification. A solar diagonal on a 75 mm (3 inch) telescope used at ×100, for example, will produce an image that is nearly 20 times brighter than the safe limit.

Additional problems with the solar diagonal arise from its use at the eyepiece end of the telescope. The objective will therefore form a real image of the Sun inside the telescope, and as in the case of projection methods, this may cause damage to the telescope structure. The wedge itself is "fail-safe" in that if it shatters from the heat from the Sun, the eye is not exposed to the full solar brightness. However, even in normal use some 90% of the solar energy passes through the diagonal and emerges from the back of the telescope. It is easy to forget this when observing, and only to remember when the smell of scorching can no longer be ignored!

If you possess a solar diagonal the best thing to do with it is to **throw it away**! Alternatively you may be able to convert it to a star diagonal for use at night, by getting the front surface of the glass wedge aluminised.

Sextants

Sextants, used by navigators to measure the position of the Sun in the sky, are specifically designed for solar observing (Fig. 2.12). They may therefore be used for viewing the Sun as well as for their normal purpose. If you have a sextant or are able to borrow one, then you may use it to observe the Sun. Most sextants include a small telescope, but the magnification is small, so that the image is similar to that for unaided observation (Chapter 6). Make sure that you follow the directions for use of the sextant supplied by its manufacturer, and it is recommended that you use all the filters that are available on the instrument. The transmission curves for the filters in a sextant are shown in Fig. 2.13. The densest of these would not be safe on its own for viewing the Sun near the zenith on a good clear day. All three filters, however, when used together do provide safe filtering.

Figure 2.12.
Observing the Sun through a sextant.

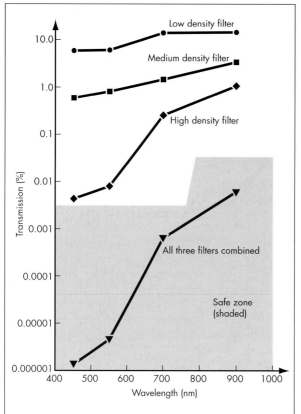

Figure 2.13.
Transmission curves for the solar filters from a sextant.

Optimising Your Telescope

Almost any design of telescope employed by amateur astronomers can be used to observe the Sun provided that it has a suitable full or subaperture filter (see the first section of this chapter) and certain designs and makes can be used to project the image, having been stopped down if need be (see the previous sections). But some telescopes and designs will produce better images and show more detail than others. Changing the design of your telescope just for solar observing is probably not an option for most people, but ensuring that the telescope that you have got performs at its best is quite straightforward. In order to see what steps need to be taken to optimise your telescope for solar observing[27] we need to see what it is about the Sun that causes problems. There are four factors to take into account:

- the enormous brightness of the Sun;
- that the Sun is a large extended object;
- some features (sunspots) are of very high contrast;
- some features (faculae, granulation) are of very low contrast.

All of the effects, however, come down to just one aspect of the telescope; its instrumental profile.[28] The instrumental profile is the image produced by the telescope of a point source such as a star (see also Chapter 4). The image of a point source, even in an optically perfect telescope, will not itself be a point. The wave nature of light ensures that through diffraction the image has a certain size and structure. That structure is shown in Fig. 2.14. The central circle will contain 60–80% of the light from the star in a good telescope. The remainder falls into the circles or fringes surrounding it. The central circle is called the Airy disk and has a diameter for visual wavelengths given by:

$$\text{Diameter of Airy disk} = 0.25/D \text{ seconds of arc}$$

---

[27] This will also optimise your telescope for most other observing programmes as well.

[28] Also known as the point spread function or PSF; see Chapter 4.

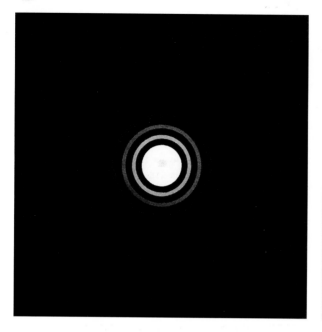

Figure 2.14. The Airy disk and surrounding fringes.

where D is the telescope objective's diameter in metres. Thus for a 0.2 m (8 inch) telescope the Airy disk is 1.25″ across.

With a good quality telescope up to 0.2 m in diameter the Airy disk and fringes can be seen directly by looking at a bright star on a night when the atmosphere is steady. They can also often be seen on photographs and CCD images. If the atmosphere is turbulent (poor seeing), the telescope not in good adjustment or not of top quality, or if it is larger than about 0.2 m, then the disk and fringes will normally be swamped by other effects, and the image of a star will be seen as a blur with a diameter at least as large as that of the Airy disk.

Now for observing stars, which are all effectively point sources, it is the size of the Airy disk, or of the equivalent blur, that governs what can be seen. The outer fringes have very little effect, and most of the time will not be seen. The resolution of a telescope is thus often quoted as half the diameter of the Airy disk (= 0.12″/D) and called its Rayleigh resolution. A good observer on a reasonable night will easily be able to see that two equally bright stars separated by the resolution of the telescope are indeed a double star. Only when the two stars are of very different brightnesses, such as

Sirius A and B which differ by a factor of ×11,400, will the fringes become of significance to the stellar observer. Then the fringes from the brighter component may swamp the fainter star.

For extended objects such as galaxies and planets which are still angularly small, the fringes of the instrumental profile become of a little more importance, but the detail seen is still largely determined by the size of the central disk. However for large extended objects such as the Sun, Moon, the Andromeda galaxy (M31), the Magellanic Clouds and the occasional comet, the situation is quite different. The level of fine detail to be seen in these objects is governed more by the outer parts of the instrumental profile than by the size of the central disk. The reason behind this is that the outer parts of the instrumental profile in large extended objects act to reduce the contrast in the fine detail, sometimes by very large amounts.

This effect may best be illustrated via an example. Suppose a telescope has an Airy disk 1″ across, which contains 70% of the light from a star, and that the outer parts (or wings) of the instrumental profile extend to 200″ and contain the remaining 30% of the light. The average intensity in the wings is thus just 0.001% that of the central disk. For planets, galaxies, small nebulae, etc., with sizes of a few tens of seconds of arc across, there will be a small but largely negligible reduction in contrast in their fine scale features.

But for the Sun the contrast change in its fine detail can be enormous. Suppose that the same telescope is now used to observe a sunspot 20″ across near the centre of the solar disk. This is easily resolved by the nominal resolution of 1″ of the telescope. The sunspot has a temperature around 4000 K and the surrounding solar photosphere one of about 6000 K. The spot has therefore an actual surface brightness about 20% that of the photosphere. But the Sun is half a degree (1800″) across and so far larger than the outermost parts of the wings of the instrumental profile of the telescope. All points within the image of the Sun will act as point sources. The light coming from each point will be spread out into the instrumental profile, with 70% going into the central disk and 30% into the wings. Thus every part of the solar photosphere within 200″ of the sunspot will spread a little light into the area covered by the spot. In total 30% of the intensity of the normal photosphere will leak into the spot. The observed brightness of the spot will thus be about

44%[29] that of the photosphere; an observed contrast of 2.3:1 instead of the true contrast of 5:1.

Thus the effect, as already stated, of the wings of the instrumental profile is to reduce the contrast within the fine detail of images of extended objects. The reduction in contrast in turn means that even features that are nominally resolved by the telescope may not be seen as separate if their initial contrast is low. Thus details of sunspots disappear or appear more muddy and blurred than they should and low contrast features like granulation may not be detected at all.

With the reason for the poor performance of a telescope in imaging the Sun identified as due to the proportion of the light going into the wings of the instrumental profile, the actions required to improve and optimise that performance become clear. That is, the telescope must be configured to concentrate the highest possible proportion of the light gathered into the central core of the instrumental profile, and the least amount possible into its wings. Even just a few of the changes and improvements suggested below can in this way result in a dramatic improvement in solar image quality.

The sources of light in the wings of the instrumental profile include:

- diffraction at the edges of lenses and mirrors and their supports;
- optical aberrations;
- misaligned/badly adjusted optics;
- poor quality optics;
- multiple reflections (especially within complex eyepieces);
- poor atmospheric seeing;
- scattering from dust, etc., on the surfaces of optics;
- scattering from imperfections within the glass of lenses and filters;
- scattering from roughnesses in the reflective coatings of mirrors;
- scattering from full aperture filters;
- scattering and reflection from structural parts of the telescope;
- scattering and reflection from the surroundings of the telescope;

and we consider how to minimise their influences below.

---

[29] $0.7 \times 20\% + 0.3 \times 100\% = 44\%$.

Diffraction

Diffraction effects cannot be completely avoided. They are intrinsic to the wave nature of light. However the more edges and supports, etc., that there are within the optical path of the telescope, the worse will be the effects of diffraction. A good refractor thus minimises diffraction, since it occurs only around the edge of the objective. A Schmidt–Cassegrain or Maksutov design is next best because although there is now a secondary mirror to add to the diffraction pattern, it is supported by the correcting lens. Finally a Newtonian or Cassegrain design will have the worst diffraction since it will occur around the supports of the secondary mirror as well as at the secondary and primary mirrors. The diffraction effects of the secondary mirror supports can often be seen on photographs of bright stars as spikes radiating out from the central image.

Home-made telescopes are sometimes provided with circular supports instead of simple radial arms for the secondary mirror (Fig. 2.15) in order to reduce the diffraction spikes. This, however, is ineffective in reducing the total diffraction contribution to wings of the instrumental profile, and so will not improve solar images.

The diffraction contribution to the wings of the instrumental profile can, however, be reduced through a technique called apodisation. This trades off an increase by a factor of two in the size of the core of the

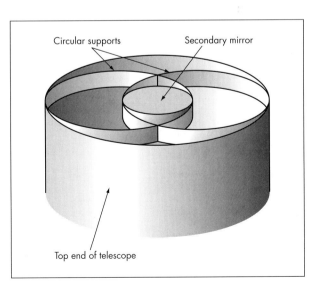

Circular supports   Secondary mirror

Top end of telescope

Figure 2.15.
Circular supports for a secondary mirror.

instrumental profile and hence a reduction in the nominal resolution, for an elimination of the fringes produced by the edges of the telescope's objective. For solar (and planetary) observation the improvement in contrast that results from the reduction in the intensity of the wings of the instrumental profile may well result in better and more detailed images, despite the poorer nominal resolution. Apodisation involves placing a neutral density filter over the telescope objective. The absorption within the filter should vary smoothly in a bell-shaped[30] manner from 100% at the outer edges to 0% at the centre. The cost of such a filter from an optical supplier would be quite horrendous, even if a supplier could be found. However almost all the improvement can be obtained using a simple home-made approximation to the filter. This requires a number of circles the size of the telescope objective to be cut from some fine mesh. The first circle then has a central hole cut in it with (say) a 20 mm (0.8 inch) diameter. The second circle then has a 40 mm hole, the third a 60 mm hole and so on, until the final hole is within 10 mm of the objective's diameter. The circles are then superimposed onto the telescope objective, making sure that the meshes are **not** aligned. If each layer of mesh allows (say) 75% of the light through, you will then have a filter whose transmission decreases from 100% over the central 10 mm of its radius through 75% from 10 mm to 20 mm, 56% from 20 mm to 30 mm, 42% from 30 mm to 40 mm, 32 % from 40 mm to 50 mm, and so on to 7% at its outer edge for a 0.2 (8 inch) telescope.

Aberration

Aberrations are defects in the images produced by optical systems. There are six main aberrations, of which spherical aberration, coma, and chromatic aberration[31] have effects which contribute to light intensities in the wings of the instrumental profile.

Chromatic aberration affects only lenses. Although it also occurs within eyepieces, the main concern is with its effects arising from the objective of a refractor. Chromatic aberration is the result of the refractive

---

[30] A Gaussian curve in mathematical terms.

[31] For more detailed descriptions of these and of the other three aberrations (astigmatism, field curvature and distortion), see for example *T&T*.

index[32] of glass varying with the wavelength of light. It results in the images in different coloured light being focused at different points. The appearance of the image of a star in a telescope suffering from chromatic aberration is thus of a sharp image surrounded by fuzzy coloured fringes. The fuzzy coloured fringes contribute to the wings of the instrumental profile of the telescope, and in extreme cases may prevent even sunspots from being seen. Any modern refractor, however, will have an achromatic lens as its objective. Such lenses combine two simple lenses so that the chromatic aberration produced by one lens counteracts that produced by the other. The effect of chromatic aberration is thus reduced, but not eliminated, through the use of an achromat for the objective of the telescope. Some modern refractor designs utilise three lenses for the objective, producing an apochromat within which the chromatic aberration is further reduced. With such a lens as its objective for all practical purposes chromatic aberration is eliminated in a refractor. Apochromatic lenses, however, are expensive and there are many refractors around which still have achromatic lenses as their objectives. In these, the residual chromatic aberration may be sufficient to cause problems with solar observing.

CAUTION
REQUIRED

Those problems can be reduced to a negligible level through the use of a colour filter; although this does of course result in a coloured image of the Sun. A colour, or band pass, filter is transparent over a small colour (wavelength) range and opaque at other colours. The filter eliminates the unfocused coloured images, leaving just the sharply focused image at its transmitted wavelength. The cheaper filters are simply dyed glass or plastic, and will usually be transparent over quite a wide range of wavelengths. Narrower band filters are available at considerably greater cost based upon the Fabry–Perot etalon (see Chapter 8). Both types of filter are available from optical suppliers, and from some telescope manufacturers. **NB** Unless the coloured filter is large enough to cover the **whole** of the telescope objective, or a stopped down part of the objective, they should **not** be used when projecting the solar image. Most coloured filters are physically small and therefore have to be used at the eyepiece end of the telescope.

---

[32] This is a measure of how much the light is "bent" by the lens. Detailed definitions and descriptions can be found in any general physics book.

They are therefore **only** suitable for use on a telescope when it already has a full aperture filter in place.

Spherical aberration and coma affect both mirrors and lenses and can come from the telescope objective and/or its eyepiece. Spherical aberration results in a sharp image surrounded by a symmetrical blur; coma in a sharp image with an asymmetrical blur. Both are severe in their effect and can reduce the proportion of light going into the core of the instrumental profile to less than 10%. Both aberrations affect all wavelengths equally, and so cannot be eliminated through the use of a colour filter.

In a reflecting telescope spherical aberration and coma can be eliminated along the optical axis by using a parabolic shape for the primary mirror. But away from the optical axis both aberrations reappear and rapidly have acute effects upon image clarity. Thus whenever observing with a reflector, the area of interest should always be moved to the centre of the field of view, so that it is on the optical axis, and unaffected directly by either spherical aberration or coma. A design of reflecting telescope which has reduced aberrations is widely used for large telescopes. This is the Ritchey–Chrétien, and it is a variant of the Cassegrain design. It uses a hyperbolic primary mirror and a stronger hyperbolic secondary mirror than the standard Cassegrain. The Ritchey–Chrétien, however, is not widely available commercially in small sizes, although some manufacturers will make one to order.

Spherical aberration and coma cannot be eliminated completely in a refractor with an achromatic objective. The effects of coma may, however, be reduced with some lens designs in which the two components of the lens are separated by an air space instead of being cemented together. In refractors with apochromats as objectives, good correction is generally achieved for on-axis images. Even with an apochromat, however, spherical aberration and coma reappear in the image away from the optical axis. It is again, therefore, best to observe the object of interest when it is in the centre of the field of view.

Unfortunately for solar work, observing objects only when they are on the optical axis will not necessarily eliminate the effects of spherical aberration and coma completely. This is because other areas of the Sun will be off the optical axis, their light will be spread into the wings of the instrumental profile and may overlap into the objects being observed on the optical axis, so reducing their contrast. The magnitude of both spherical aberration and coma, however, increases inversely with the square of the focal ratio of the telescope.[33] A considerable reduction in their effects and improvement in the overall image quality, at a cost of a worsening of the nominal resolution, may therefore be achieved by stopping down the telescope. Stopping down a telescope will increase its focal ratio. The effects of coma and spherical aberration will thus be reduced by a factor of 4 if the objective is stopped down to half its normal diameter.

Optics

The optics of a telescope may be out of adjustment or misaligned in several ways, but all with deleterious effects upon the image. You should consult your telescope handbook and if need be the manufacturer on how to check the alignment of its optics, and how to adjust the alignment if required. With many designs, however, especially the Schmidt–Cassegrain and Maksutov, the realignment will require specially made jigs, and so can only be carried out by the manufacturers or their agents. It should also be borne in mind that it is possible, especially for the larger, heavier, and/or home-made instruments, for a telescope to be perfectly aligned when pointing at one part of the sky, but then to go out of alignment as it is moved to point to another part of the sky. The change arises through the telescope structure flexing under the changing gravitational loads as it is moved on its mounting. Telescopes on alt-azimuth mountings will generally be less affected by flexure, since the telescope only changes its attitude in the vertical plane.

There is an additional aspect to misalignment that affects Cassegrain, Ritchey–Chrétien, Schmidt–Cassegrain and Maksutov designs, and which is rarely realised by their owners. With these designs, the focusing is often achieved by changing the separation of the primary and secondary mirrors. This moves the focus position towards or away from the back of the primary mirror. However, there is only one separation

---

[33] For designs like the Schmidt-Cassegrain and Maksutov, it is the final focal ratio that is important here, not that of the primary mirror, which has normally a much smaller value.

for the mirrors which is correct in the sense that the aberrations are minimised or eliminated.[34] Changing the separation from that ideal value, even if in all other respects the telescope is perfectly aligned, rapidly leads to the introduction of spherical aberration even for on-axis images. The telescope should thus be used with the mirrors correctly separated wherever possible. Since telescope handbooks rarely mention this problem, you will probably need to contact the manufacturer to get the information required to set up the telescope correctly.

Poor quality optics should not be a problem except with the very cheapest of instruments. However the normal quality of a good telescope may not be adequate for solar observing. Assuming that the telescope has been designed properly, then the quality of its performance is determined by the accuracy with which the optical surfaces can be manufactured to fit the design specifications. The commonly quoted accuracy that is needed for mirrors is one eighth of the operating wavelength.[35] For visual work this is ±60 nm (±2.5 millionths of an inch). Such a required level of precision for the optical surfaces, however, is based upon reaching the Rayleigh resolution (see above) of the telescope when observing double stars. That in turn is dependent upon the core of the instrumental profile. But as we have seen it is the wings of the instrumental profile that have a significant effect upon solar image quality. The wings of the instrumental profile can continue to be improved, by reducing their extent and the proportion of light going into them, if the optical surfaces are produced to an accuracy better than one-eighth of a wavelength. In an ideal solar telescope, the optics should be accurate to a twentieth

---

[34] For example with the Cassegrain design, the back focus of the hyperbolic secondary mirror must coincide with the focus of the parabolic primary mirror.

[35] The required accuracy is sometimes expressed in terms of deviation from the ideal of the shape of the wavefront of the light beam. The required accuracy is then a quarter of the operating wavelength for the wavefront, but this still translates to an eighth wavelength for the optical surfaces' accuracy. The surfaces of the objective lens in a refractor need only be accurate to about a quarter of the wavelength, since the focusing effect of the lens is spread over at least two surfaces.

or better of the operating wavelength. Few telescope manufacturers produce instruments to this level of surface accuracy, and if they were to do so it would probably be at prohibitive cost. Perhaps surprisingly, though, it is entirely possible for an experienced amateur telescope maker, used to grinding and polishing his/her own optics, to work to twentieth wave accuracy and so to produce an optimised solar telescope. Such an instrument would have the advantage of also giving excellent results for planetary work.

Almost all reflecting telescope designs use multiple reflections to produce their final image and these are obviously desirable. However plain glass surfaces reflect about 5% of the light falling on them (see the subsection on reflection). So the lenses in a refractor and in eyepieces can have reflections occurring between the plain glass surfaces involved, and these are not desirable. While antireflection coatings can be used, these only completely suppress the reflection at one wavelength. The multiple reflections within and between lenses, or from lens to mirror surfaces and back, usually produce faint secondary images some distance away from the main image. This is not a great problem when observing stars, planets, etc., since the secondary images are well separated from the main image. But for solar work, the secondary images of parts of the Sun well away from the optical axis may well overlap the main image on the optical axis. If it is the main components of the telescope that are involved in the multiple reflections then there is little to be done about it. Eyepieces, however, can easily be interchanged, and one with fewer or no multiple reflections used for solar work. Generally the more complex and expensive an eyepiece is, the more likely it is to have multiple reflections, since such eyepieces use several lenses with air gaps between them. This type of eyepiece should therefore be avoided for solar work. The monocentric eyepiece design usually has minimal multiple reflections, since its components are cemented together. Any eyepiece can be checked for multiple reflections by using it to observe a bright star or planet. The image is first centred in the field of view, and then allowed to drift to and over the edge. Multiple reflections will appear as fainter images, often out of focus, which move in a different direction from the drift motion of the real stars in the field.

Seeing

The blurring effect of turbulence in the Earth's atmosphere cannot be completely avoided except by going into space or sometimes through the use of highly sophisticated and expensive adaptive optics systems. Steps can, however, be taken to reduce atmospheric effects. For solar observing the most effective single action is to observe early in the morning. Between one and two hours after dawn, the Sun is high enough in the sky to be above the thickest parts of the atmosphere, but has not yet had time to create further turbulence by heating the Earth's surface and atmosphere.

For a skilled observer, better images may be seen than recorded by a camera (Chapter 4). This is because the turbulence varies in its effects on the image. For brief intervals, the atmosphere may be still enough to allow through an undistorted image. The visual observer can learn to see the fine details in those one or two seconds of clarity, and to ignore the blurred images available the rest of the time. Obtaining a photographic or CCD image (Chapter 4) during one of these clear moments would arise only through chance.

A high proportion of the atmospheric turbulence occurs within a few tens of metres of the ground and arises through local effects. Several steps can therefore be taken to reduce ground level turbulence. One is to observe from an area likely to be at or near the ambient temperature of the atmosphere, so that thermal currents are avoided. Suitable ground cover for this would be a large lawn with grass long enough to cover the bare earth, an area of low-growing bushes, or an island in the middle of a lake. Areas covered in concrete or tarmac, etc., should be avoided. Similarly, buildings, isolated trees, hills, valleys, etc., where wind is likely to produce swirls and eddies should be avoided, although a line of trees some distance away to act as a wind break can be useful. If a building does not itself create turbulence or thermal currents, then observing from its roof, if this can be done in safety, will lift the telescope above much of the ground level turbulence. Fine mesh screens (not solid screens, since these can create eddies) can be used in the immediate area of the telescope or observatory to reduce turbulence from wind, and these will also reduce wind shake on the telescope itself.

Stray Light

Scattered and reflected light can be minimised by keeping all optical surfaces clean and dust free. However, care is needed in cleaning optical surfaces, and proprietary cleansers (advertised in the popular astronomy magazines) only should be used and their instructions followed, otherwise coatings and even the glass surfaces themselves can be damaged. The reflective coatings on mirrors should be renewed regularly, and certainly long before they have notice-ably deteriorated. Imperfections in the glass of lenses, etc., should not occur in good modern telescopes, and if they do, then the instrument should be returned to the manufacturer for replacement under its guarantee. However, in older instruments there may be bubbles or other imperfections in the glass, or you may have scratched one of the optical surfaces in a more modern instrument. Scattering from scratches and bubbles, etc., can be reduced by making a cardboard mask to cover the affected part of the lens or mirror but do not be tempted to stick a bit of tape, etc., on the surface; this will almost certainly damage any coatings there may be on the optics.

All internal parts of the telescope, other than the optics, should be painted matt black to reduce reflections. With a modern commercially produced telescope, this will almost certainly already be the case. If not, then the telescope should be returned to the manufacturer for any uncoated surfaces to be painted black. Do not attempt to do this yourself, since dismantling and painting a telescope internally will certainly invalidate its guarantee, and with some designs the reassembly and realignment may only be possible using special jigs.

If your telescope has an open structure for its tube, then this should be boxed in to prevent light entering the telescope from any direction except the desired one. Often this can be done on a temporary basis just by taping a sheet of thin black card around the tube. Finally the telescope should be used under as dark conditions as possible. If the telescope is in the open, then there is probably little that can be done, but if it is in an observatory, then the dome aperture should be opened to the minimum required to allow the solar light to reach the telescope. Even better, the dome aperture should be covered completely except for a hole the size of the telescope objective, through which the

Sun is observed. In the latter case the aperture mask will need to be moved frequently to follow the motion of the Sun across the sky. One way of accomplishing this simply is to use a mask with a hole several times the size of the telescope objective, and then to cover that hole with a second mask held in place with Velcro strips, which has the telescope-sized hole cut into it. The second mask can then quickly be repositioned as needed.

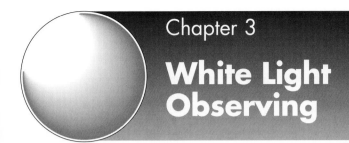

**CAUTION
REQUIRED**

Note the warning at the start of this book, and do not attempt any solar observation until you have assimilated all the details in this and other chapters.

This chapter deals with the specific requirements for observing individual types of solar feature. The general requirements for solar observing and for optimising a telescope for this type of work are covered in Chapter 2.

Sunspots

Sunspots (Chapter 1) are very dark compared with the bright solar photosphere. Their presence is therefore easily detected via any safe method of observing the Sun, and few or no extra adjustments need to be made. Determining the sunspot number or following the rotation of the Sun (Chapters 1 and 5) can therefore be accomplished in a straightforward manner with any instrument. The smallest sunspots are only a few seconds of arc across and will need magnifications of ×150 or more to be seen with certainty. Most spots and groups, however, are considerably larger than this and can be seen at magnifications which allow the whole disk of the Sun to be viewed (typically ×100 or less). Thus the movement, pattern and distribution of spots across the Sun's disk can be followed.

Seeing the details of sunspots, however, is a different matter. Structures within spots occur on scales from tens of arc seconds down to fractions of an arc second. This corresponds to physical sizes from 10,000s to 100s of kilometres. The umbra of a sunspot is generally featureless, although very, very occasionally traces of faint, extremely low contrast patterns similar to granulation may be glimpsed within it, especially when it is near to the centre of the solar disk. The penumbra usually has a radial structure with alternating dark and light lines. These have a reasonable level of contrast both between themselves and with the umbra of the spot and the solar photosphere. A good deal of structure can therefore be seen in spots, especially within the penumbra at magnifications of ×150 or so, with a few minimal steps, such as cleaning the optics (Chapter 2), taken to ensure that the telescope is performing well. For the finest, low contrast detail, magnifications of ×300 or greater will be needed, and observations will have to be made under the best possible observing conditions and with a telescope optimised in all respects for solar work (as explained in Chapter 2).

More detail may sometimes be seen when the spot is near to the limb of the Sun. Although the spot is then seen at an angle and is foreshortened, limb darkening reduces the surrounding photospheric intensity to 40% of that near the centre of the solar disk. Also half the field of view is dark. The contrast-reducing effects of the wings of the instrumental profile of the telescope (Chapter 2) are thus considerably ameliorated. Observations of spots near the solar limb are also needed to pick up the Wilson effect (Chapter 1).

Sunspot groups are composed of individual spots, though often with some of them merging and overlapping. The requirements for observing the overall development of a group are therefore similar to observing the grosser structures within individual spots. Similarly the fine detail of groups has similar observing requirements to that for the fine detail of spots. Very occasionally a solar flare (Chapters 1 and 8) may be intense enough to be observed in white light. Unless you have a radio telescope (Chapter 9) to give warning of the occurrence of a flare, seeing one in white light is largely a matter of chance. Flares, however, occur mostly during the earlier stages of very large complex sunspot groups. Observing such groups during their development therefore gives the best chance of seeing a flare. In white light, flares generally appear as low contrast small brightenings within and around the sunspot group. Any steps taken to reduce the wings of the instrumental profile of the telescope (Chapter 2), will therefore help with flare observation.

Limb Darkening

Limb darkening (Chapter 1) causes the edge of the solar disk to have an intensity in the visual region of only about 40% that of its centre. It is best observed using a magnification which enables the whole solar disk to be seen. This means ×100 or less for most eyepiece designs. Using a full aperture filter, limb darkening will normally then easily be seen without any other special adjustments being needed to the telescope. When projecting the solar image, a good shade will normally be required, or the scattered solar light from the surroundings will make the limb darkening difficult to distinguish.

With some telescope and eyepiece designs, vig-netting[36] may mimic solar limb darkening. Vignetting is usually symmetrical in its effects over the field of view. So if you suspect its presence, then limb darkening can still be observed by using a very low magnification (×50 or less), and positioning the solar image towards the edge of the field of view. The limb darkening will then be symmetrical about the solar image while the vignetting will cause a gradual shading across the whole disk. Alternatively, the instrument can be checked for vignetting by observing the full moon. Since the Moon does not have limb darkening, if a similar shading towards its edge is observed, then this must be due to vignetting.

Granulation

Granulation (Chapter 1) is best seen near the centre of the solar disk. It is small in scale (1–2″) and of low contrast. It therefore needs a magnification of ×150 or more, and at least some of the simpler steps, such as cleaning the optics (Chapter 4), taken to ensure that the telescope is performing well. It is also advantageous to be observing under good atmospheric conditions. It will then be seen as a mottled pattern covering the whole photosphere, except where there are sunspots. The

---

[36] This is a shadowing of the field of view caused by stops, secondary mirrors, undersize optics or other physical structures, such as mirror supports, within the optical path of the telescope.

individual mottles have a lifetime of about 10 minutes, and so their development can readily be followed. Since the dark lanes between the mottles are slightly cooler, and therefore redder, than the centres of the mottles, the visibility of granulation can be enhanced through the use of a green-blue or yellow-green transmission filter. This increases the contrast in the image by darkening further the dark intermottle lanes while allowing through the light from the centres of the mottles.

Faculae

Faculae (Chapter 1) are of low contrast and individually are just a few arc seconds in size. They are associated with sunspots and usually occur within a few tens of arc seconds of the spot. They appear before the spot and remain after it has disappeared, so they may also be seen in isolation. There are usually many faculae around a sunspot or sunspot group and so the total area covered by them may extend over tens or hundreds of seconds of arc. They are best observed when the spot is near to the limb of the solar disk. Limb darkening then enhances their visibility by improving their contrast with the photosphere. Observation of faculae requires magnifications of ×100 or so if the whole area is to be seen, and of over ×150 for their details. Following some or all of the precautions to ensure that the telescope is optimised for solar work (Chapter 2) will greatly improve the visibility of faculae.

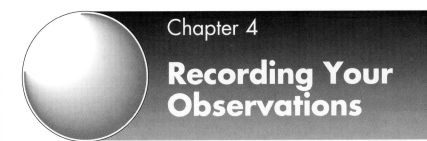

Chapter 4

Recording Your Observations

The solar images produced by projection or through full aperture filters can be recorded by drawing or by using photographic, CCD,[37] digital or video cameras. Whatever device is used to record the solar image, following as many as possible of the procedures described in Chapter 2 to optimise that image will be very beneficial.

CAUTION
REQUIRED

Make sure that you have assimilated Chapter 2 and the warnings therein, especially on some manufacturers' restrictions on the use of some telescopes for projecting the solar image, before attempting any procedures discussed in this chapter.

Drawing

Some practical ability is involved in any type of telescopic work, but making accurate drawings requires skills which are not automatically associated with being an astronomer. None the less, practice and a few tricks of the trade should enable anyone to make usable sketches. While making drawings may sometimes be regarded as "second class" compared with obtaining any type of camera images, this is not always true. A skilled observer can see as much as the camera can record, and can often see more. The reason for this lies in the atmospheric turbulence which blurs images (see also Chapter 2). The turbulence (which results in

---

[37] Charge coupled device.

scintillation, seeing or twinkling of images) varies on a time scale of a few milliseconds. Occasionally, there may be periods of a few seconds when the atmosphere along the line of sight from the telescope to the Sun just happens to be still and uniform. The detail visible within the image will then be determined by the resolution of the telescope and not by the blurring of the atmosphere. The visual observer can learn to make his/her observations just during those instants of clear seeing, and then to record the details onto a drawing. Some care, however, needs to be exercised in this process, since very fine detail may be glimpsed, but not seen correctly – the Martian canals arose from this type of mistake.

It is probably easiest to make drawings from projected (Chapter 2) images because the main features can be sketched directly onto paper attached to the projection screen. A firmly mounted and driven telescope will be needed to do this otherwise the act of drawing will move the image. If you sketch the Sun regularly using the same instrumental set-up, then you can pre-prepare drawing paper with an outline of the Sun, and align the actual image onto the drawn circle using the telescope's slow motions. The details of sunspots and sunspot groups, etc., can be put onto the outline sketch after it has been removed from the projection screen. Alternatively grids can be drawn onto the projection screen and the drawing paper. The grids are then used to determine accurate positions for the solar features seen on the screen and to transfer them to the drawing.

Making a sketch from a view through an eyepiece is considerably more difficult to do accurately than from the projected image. It is easy to exaggerate the details and so to distort the overall image. It is best to use a wide angle eyepiece to obtain the main features of the image first, and then to fill in the details using higher magnifications. A pre-prepared outline of the Sun and a grid pattern on the drawing paper will again help to give precise images. If you have a cross-wire eyepiece or a bi-filar micrometer[38] (but note the warning on **not** using these for eyepiece projection, Chapter 2), then

---

[38] This is an eyepiece with one fixed cross-wire, and a second single wire which is parallel to one of the cross-wires and may be moved along the line of the other cross-wire. The position of the moveable wire is given by a vernier or direct read-out display on the screw drive for the wire. The whole device may be rotated and its position angle read from a circular scale

precise relative positions for objects may be measured and transferred to the drawing. The micrometer may be used to get the relative positions in two dimensions for any mutual orientation of two features, or to measure the two-dimensional shape of a single feature. The fixed cross-wire eyepiece can give positions and sizes directly only along one direction. An accurate scale in one direction, however, will help you to judge positions and sizes perpendicular to that direction. Positions and sizes are measured using a fixed cross-wire by turning off the telescope's drive and allowing the image to drift across the vertical wire. The time interval T (in seconds of time) for a feature to drift across the wire, or for the drift between two features, is then related to their angular separation along the drift direction by

Angular size or separation $= 15 \cos \delta \times T$ seconds of arc where δ is the declination of the Sun.

Whether drawing from a projected image or from looking through an eyepiece, very fine details may sometimes be more easily seen if the telescope is moved very slightly. This can be done just by nudging the instrument or by small movements using the slow motions. The human eye–brain combination has evolved to be very sensitive to small movements,[39] and so fine detail in the image becomes more easily seen when it is moving slowly.

Photography

For many purposes photographic and digital cameras are interchangeable. Digital cameras have the advantage of producing images which may be loaded directly into a computer for image processing (discussed later in this chapter), but generally have poorer angular resolution. They are quite sophisticated devices with autofocusing

---

(footnote 38 continued)
around it. It is used by centring the fixed cross-wire on one feature, rotating the micrometer until the fixed wire bisects it, and positioning the moveable cross-wire on the centre of the second feature. The relative positions of the two features are then obtained from the micrometer's position and position angle scales. See *AT* for further details.

[39] Presumably so that our ancestors avoided sabre-toothed tigers better.

and autoexposure usually as standard features. However, autofocusing may cause problems when using the camera on the telescope since the camera may attempt to focus on parts of the telescope structure rather than the image or be upset if polarisation is introduced into the image by the full aperture filter (Chapter 2). The same comment applies to autofocusing photographic cameras. Thus a relatively simple SLR photographic camera with manual focusing and exposure settings may be better for solar imaging (and also for more conventional astronomical photography). Such cameras can often be picked up second hand very cheaply, and so having one dedicated for use on the telescope is not an unreasonable option. The following sections give observing procedures based around manual photographic cameras, but where their additional features do not cause problems, digital or more sophisticated photographic cameras may be used instead.

Almost any film type – fast or slow, fine grained or coarse grained, colour or monochrome (black and white), print or slide – can be used for imaging the Sun. If using a full aperture filter with a colour bias (Chapter 2), then you can with advantage use monochrome film, since the colour in the image is spurious anyway. Fast film will enable shorter exposures to be used, with less chance of atmospheric blurring, wind shake or tracking errors becoming apparent. But fast film is also coarse grained, and so images on it cannot be enlarged as satisfactorily as those on fine grained emulsion. For real colour images, and especially of the corona and prominences during total eclipses (Chapter 7), colour slide film gives richer colours than colour print film. However, slide film has less latitude for exposure errors than print film. Finding the best film for your purposes requires trading off the advantages and disadvantages of each type, and probably ultimately trying out several types until you find which gives you the best results. The answer will also undoubtedly change should you decide to image faculae rather than sunspots, Hα features rather than total eclipses, granulation rather than limb darkening, and so on.

Projected Images

The projected image is most simply recorded by pointing the camera directly at the screen. Normally

the camera exposure meter will provide the exposure
that is needed, although it is always a good idea to try
exposures on either side of the one recommended (i.e.
if a photographic camera suggests an exposure of, say,
1/125 s, try also 1/250 s and 1/60 s, and even 1/500 s and
1/30 s). Since the camera will be placed to the side of the
telescope, there will be some distortion of the image
because the viewing angle will foreshorten the image
along the axis containing the camera and the telescope,
but not the axis perpendicular to that. For non-critical
work this may not matter since the distortion is
normally quite small. If needed, however, the screen
can be tilted at an angle half way between the telescope
and the camera (Fig. 2.9). Normally there will be
sufficient depth of focus in the projected image from
the telescope for it to remain in focus across the whole
disk. The tilted screen will distort the image in the
opposite sense to the foreshortening of the camera
viewing angle, and the final image will then be correct
(Fig. 4.1). With a CCD image (discussed later) or a
photographic image that has been scanned into a
computer, the distortion can easily be corrected by
image processing (described later in this chapter).

Full Aperture Filters

Photography of the Sun using a metal-on-plastic full
aperture filter is similar in principle to imaging objects
at night. The telescope is either used as a very long focal
length telephoto lens in place of the camera's normal
lens, or an eyepiece is also used to give an enlarged
image. If you are already experienced at astronomical
photography, then your normal methods should work
for solar imaging. The exposure times, however, will be
much less than those to which you are probably
accustomed. Table 4.1 gives a guide to the required
exposures when using a typical double layer filter. Since
filters differ, however, it will be necessary to calibrate
the particular telescope–filter–camera–emulsion type
system that you are using. Once calibrated, Table 4.1
should thereafter enable you to estimate suitable
exposures when you change the film speed or projection distance, etc. Brighter images may be obtained with
single layer filters (such as metal-on-glass filters), and
have the advantage of reduced exposure times.

To use your telescope as a long focal length telephoto
lens, then your camera must be of the single lens reflex

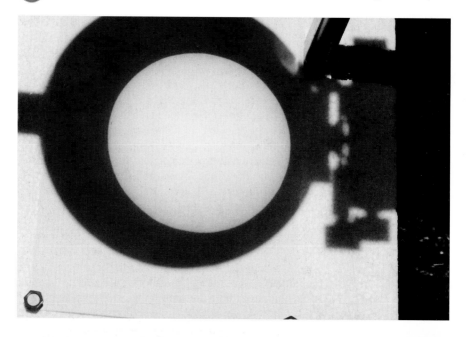

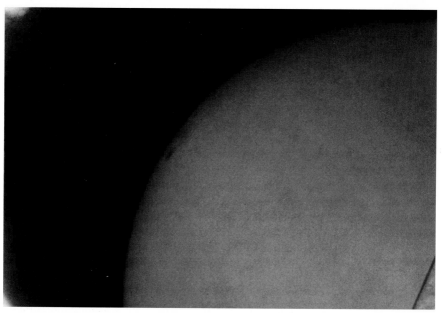

Figure 4.1. Photographic images of the Sun obtained by eyepiece projection. Top: the whole disk showing distortion due to not using a tilted screen; bottom: an enlarged image using a smaller focal length eyepiece and with the geometrical distortion corrected through the use of a tilted screen.

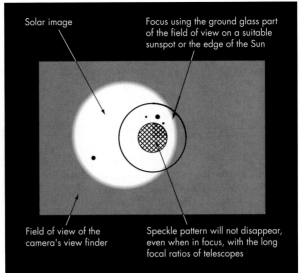

Solar image

Focus using the ground glass part of the field of view on a suitable sunspot or the edge of the Sun

Field of view of the camera's view finder

Speckle pattern will not disappear, even when in focus, with the long focal ratios of telescopes

Figure 4.2. Using the ground-glass screen area of an SLR camera's viewfinder to focus a solar image.

(SLR) design,[40] so that the viewfinder will look through the telescope when the camera is attached to it. It is also better if the camera is of a simple manual design; autofocusing and autoexposure cameras will often fail to function properly when the camera is on the telescope. The normal lens of the camera must be removed, and then with the filter **still in place** on the telescope, the camera is attached to it using a suitable adaptor[41] in place of the eyepiece. The image of the Sun may be seen through the viewfinder and focused and centred as required.

NB Many cameras have optical devices to aid focusing. These often show a speckle pattern in a central circle within the viewfinder when the camera is not in focus. These focusing devices will not work with the long focal ratios provided by most telescopes. Instead you will need to focus within the ground-glass screen part of the viewfinder. This is often present as an annulus around the centre circle (Fig. 4.2), or sometimes may form the whole of the outer part of the field

---

[40] If your telescope has a precise and reliable position indicator on its focusing mechanism, then it is possible to use non-SLR cameras. You will need to take a large number of photographs over a range of focus settings, however, in order to find the correct position. Even then, you will not be able to position the image where you want it on the negative.
[41] Usually available for most makes of camera from the manufacturer of your telescope.

of view of the viewfinder. Some manufacturers produce optional magnifiers for their camera viewfinders. If this is the case for your camera, then it will usually be worthwhile obtaining one for use both for solar work and for night-time astronomical imaging. Whatever your camera and its set-up you will almost certainly need to experiment to find how best to focus, before taking any images.

With the full aperture filter in place, the solar image may be difficult to see in the camera viewfinder, because of the bright conditions in which solar observing occurs. This will especially be the case at long focal ratios and with magnified imaging (discussed below). A single layer filter, such as a metal-on-glass filter, may give a brighter image which is more easily visible. Failing that, the ambient light level will need to be reduced so that the viewfinder image has an improved contrast. In an observatory, this can be done by narrowing the dome slot until the aperture is small enough that only the telescope is illuminated by the Sun. Alternatively (or additionally), a sheet of dark cloth can be draped over your head, the camera and the eyepiece end of the telescope, to provide a shield. If focusing still remains difficult, then you may need to undertake trial exposures at noted focus settings, develop the film, and select the setting which gives the best focus for future use. You may also be able to focus on a night-time object and note the focus setting or leave the system set up for solar work during the daytime.

Providing that the solar image is large enough to cover the detector of the camera's exposure meter, then this may be used to set the exposure. Alternatively Table 4.1 may be used to estimate the required exposure time. Remember that if the "full" aperture filter does not in fact cover the whole aperture, then the actual clear area of the filter will be needed to calculate the focal ratio.[42] Since filters, film, telescopes and cameras all vary one from another, the figures in Table 4.1 are only a guide. For your first attempt at solar photography you should use a wide range of exposure times. The best of those exposures can then be used to

---

[42] For the set-up shown in Fig. 2.3, the telescope has an original focal ratio of f10, but with the 125 mm filter in place plus the obscuration by the secondary mirror, it is actually working at f18, leading to suggested exposures around 1/30 to 1/15 of a second from Table 4.1.

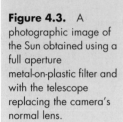

Figure 4.3. A photographic image of the Sun obtained using a full aperture metal-on-plastic filter and with the telescope replacing the camera's normal lens.

calibrate your system for future use. For example if the table suggests 1/30 of a second exposure time, and your best image was obtained at 1/125 of a second, then the same factor of four (two photographic stops) should be used when you change the film speed or try eyepiece projection, etc. Photographs of the Sun, comparable in quality to those you obtain with night-time observations, should be attainable (Fig. 4.3).

CAUTION
REQUIRED

Larger scale solar images can be obtained by eyepiece projection, just as large scale images of the Moon or planets can be obtained at night. But note: this is **not** the same eyepiece projection method that has been discussed earlier. Here, the full aperture filter **must** be in place on the telescope. The camera and eyepiece are then set up for projection in the same manner as for night-time observations (Fig. 4.4). The same type of difficulties will be experienced as for those night-time observations, especially the difficulty in getting precisely focused images. Beware also of any dust on the lenses of the eyepiece, since this will project as blemishes onto the image (Fig. 4.5). The required exposure can be estimated from Table 4.1, but you will need to calculate the effective focal ratio. This is given by:

$$\text{Effective focal ratio} = \frac{\text{Original focal ratio} \times \text{Projection distance}}{\text{Eyepiece focal length}}$$

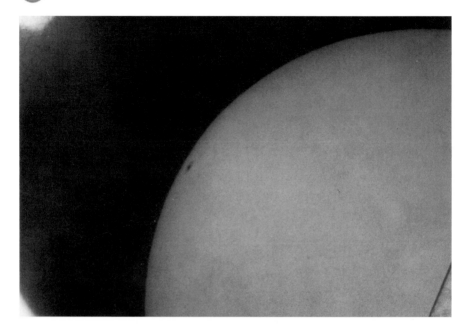

Figure 4.4. Eyepiece projection imaging of the limb sunspot in Fig. 4.3 using a full aperture filter. Note the increased magnification.

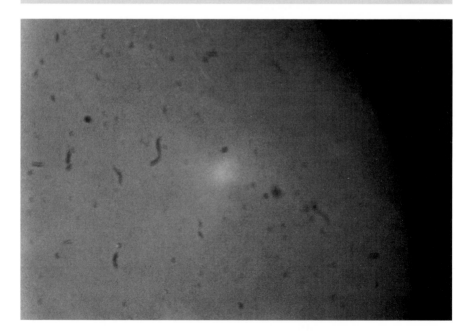

Figure 4.5. How not to do it! An eyepiece projection image of the Sun through a full aperture filter. There are two sunspots just to the right of the centre; the remaining features are all dust particles on the eyepiece.

Table 4.1. Exposure guide to the photographic imaging of the Sun through a full aperture double metal-on-plastic solar filter

Focal ratio or effective focal ratio	Exposure time[a] for ISO 100/21° film (seconds)	Nearest preset exposure time(s) on most cameras
4	0.001–0.005	1/1000, 1/500, 1/250
6	0.002–0.01	1/500, 1/250, 1/125
8	0.005–0.02	1/250, 1/125, 1/60
10	0.008–0.03	1/125, 1/60, 1/30
12	0.01–0.05	1/125, 1/60, 1/30. 1/15
14	0.015–0.06	1/60, 1/30, 1/15
16	0.02–0.08	1/60, 1/30, 1/15, 1/8
18	0.025–0.1	1/60, 1/30, 1/15, 1/8
20	0.03–0.1	1/30, 1/15, 1/8
30	0.07–0.3	1/15, 1/8, 1/4, 1/2
40	0.1–0.5	1/15, 1/8, 1/4, 1/2
50	0.2–0.8	1/8, 1/4, 1/2, 1
75	0.4–2	1/2, 1, time exposure 2 s
100	0.8–3	1, time exposure 3 s
150	2–7	time exposure 2–7 s
200	3–12	time exposure 3–12 s

[a]For other film speeds, adjust the timings in proportion to the ratio of the first parts of the two ISO numbers; i.e. for ISO 50/18°, the times must be doubled (100/50 = 2), for ISO 200/24° they must be halved (100/200 = 0.5), for ISO 400/27° they must be quartered (100/400 = 0.25) and so on.

The original focal ratio used in this equation will need to take account of the actual clear area of the filter if it is smaller than the whole aperture of the telescope. Also, as before, you should always try a range of exposure times. Though if you have already calibrated your system with a different set-up, the calibration factor can be used in conjunction with Table 4.1 to get a more reliable initial estimate of the time required.

The size of the solar image can be found from Table 4.2, using the actual or effective focal length of the telescope. The effective focal length is just the effective focal ratio multiplied by the diameter of the telescope, or the effective diameter if using a sub-aperture filter. Thus for a telescope with a 0.125 m (5 inch) full aperture filter and with a focal length of 1.8 m (71 in), the solar image would be about 15 mm (0.6 in) across. If the same instrument were then to be used for eyepiece projection with an eyepiece of 40 mm (1.6 in) focal length, and a projection distance of 100 mm (4 in), then from the above equation it would

Table 4.2. The size of the solar image

Focal length or effective focal length (metres)	Solar image diameter (mm)	Focal length or effective focal length (feet)	Solar image diameter (in)
0.5	4.4	1	0.10
1.0	8.7	2	0.21
1.5	13.1	3	0.31
2.0	17.5	5	0.52
3.0	26.2	10	1.05
5.0	43.6	15	1.57
10.0	87.3	30	3.14

have an effective focal ratio of f36.[43] The effective focal length would thus be 4.5 m (15 feet),[44] and so from Table 4.2, the image of the Sun would be about 40 mm (1.6 in) across.

The exposure time (from Table 4.1) for ISO100/21° film would be around 1/60 to 1/15 of a second for the direct image at f14.4 through the full aperture filter and about 1/8 to 1/4 of a second for the projected image at f36 through the same filter.

The exposures needed for the Sun are often short enough that the telescope need not track the Sun during the exposure. The Sun moves through its own diameter in about 2 minutes. Photographic emulsion is typically able to resolve between 40 and 100 lines per mm depending upon whether it is fast and coarse-grained or slow and fine-grained. The image will not be blurred by the Sun's motion if during the exposure it does not move more than the resolution limit of the emulsion. Thus the above 40 mm (1.6 inch) solar image will move through its own diameter in 2 minutes, and it is therefore moving at a rate of 0.33 mm/s (0.013 inches/s). With fine grained film, the resolution is about 0.01 mm. Thus an exposure shorter than 1/30 second (0.01/0.33) will not show any image movement. The low angular resolution of digital cameras means that unblurred exposures at least as long as those for fast film may be used. Table 4.3 lists the longest such non-blurring exposures for a range of focal lengths.

If you have an Hα filter (Chapter 8), then images can be obtained using the same methods as for the full aperture filters (Fig. 4.6). However, since the Hα line is at

---

[43] f14.4 × 100 / 40 = f36.
[44] Filter diameter is 125 mm, so 125 × 36 = 4500 mm = 4.5 m.

Table 4.3. Maximum exposure times for an undriven telescope if the solar image is not to be blurred

Focal length (m)	(ft)	Maximum non-blurring exposure (s) Fast film and digital cameras		Slow film	
0.5	1.7	0.60	($\simeq$1/2)	0.28	($\simeq$1/4)
1.0	3.3	0.30	($\simeq$1/4)	0.14	($\simeq$1/8)
1.5	5	0.20	($\simeq$1/4)	0.09	($\simeq$1/8)
2.0	6.7	0.15	($\simeq$1/8)	0.07	($\simeq$1/15)
3.0	10	0.10	($\simeq$1/8)	0.05	($\simeq$1/15)
5.0	16	0.06	($\simeq$1/15)	0.03	($\simeq$1/30)
10	30	0.03	($\simeq$1/30)	0.014	($\simeq$1/60)

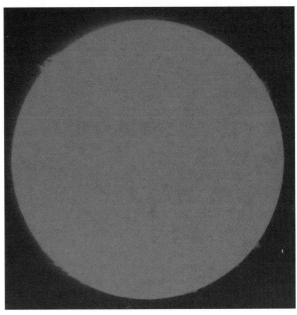

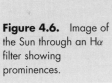

Figure 4.6. Image of the Sun through an Hα filter showing prominences.

a wavelength of 656 nm, in the deep red, you will need to make sure that the photographic film that you use is sensitive to that part of the spectrum. Most colour films will work, but many black and white films will cut off at shorter wavelengths. In either case seek the advice of the retailer or manufacturer before using the film.

Magnified Imaging

An alternative projection method, which is in some ways simpler and easier than the above, and which can

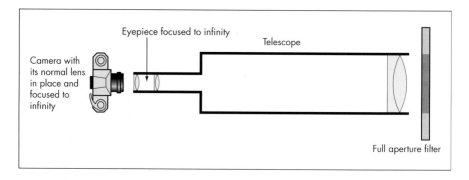

Eyepiece focused to infinity

Telescope

Camera with its normal lens in place and focused to infinity

Full aperture filter

be used with any camera not just SLR designs, is termed magnified imaging.[45] In effect the telescope or binoculars with a full aperture filter and an eyepiece in place produces a magnified version of Sun, and the camera is used with its normal lens in place to photograph this (Fig. 4.7). Autofocusing cameras, however, will generally not work, since they will attempt to focus on the eyepiece or telescope and not on the image. If you are using a Mylar-based full aperture filter, then this may introduce some polarisation into the solar image. Such polarisation can also lead to autofocus systems failing to work properly (Chapter 2).

In magnified imaging the telescope is focused to infinity, i.e. so that its image would be in focus for an eye viewing a distant object. This is the normal setting for many people, but if you are short sighted, then you will need to have your glasses on when focusing the telescope (see below for more details on this). With the telescope focused on infinity, the light emerges from the eyepiece as parallel beams and will not produce a real image if you try to project it onto a screen. The camera is placed behind the eyepiece with its lens also set to focus at infinity. If the two focal settings are accurate, then there will be an in-focus image on the photographic emulsion. The size of the image will be increased compared with a direct image taken using the same camera and lens by the magnification of the telescope.

The exposure that is needed can be calculated by determining the effective focal ratio of the telescope and camera combination. The effective focal length is the focal length of the **camera lens** (not the telescope),

Figure 4.7.
Magnified imaging of the Sun.

---

[45] Also known as afocal imaging.

multiplied by the **telescope** magnification. The effective focal ratio is thus this effective focal length divided by the diameter of the **telescope objective** (not the camera lens), or of the full aperture filter if this is actually smaller than the objective:

$$\text{Effective focal ratio} =$$

$$\frac{\text{Focal length of camera lens} \times \text{Telescope magnification}}{\text{Telescope objective (or filter) diameter}}$$

Thus using a 40 mm eyepiece on a 0.2 m f10 Schmidt–Cassegrain telescope would give a magnification of ×50. With a camera with a 50 mm (2 inch) lens the effective focal length would be 2500 mm (= 50 mm × 50, or 2.5 m, or 100 inches). From Table 4.2 the solar image would thus be 22 mm (0.9 inches) across. The effective focal ratio would be f12.5 (= 50 mm × 50/200 mm), and so from Table 4.1, the exposure required would be about 1/60 or 1/30 second.

With a full aperture filter on a pair of 7 × 50 binoculars[46] and a camera with a 135 mm telephoto lens, the effective focal ratio would be f19 (= 135 mm × 7/50 mm), and the exposure needed about 1/15 or 1/8 seconds. The effective focal length would be 0.945 m (= 135 mm × 7), and so from Table 4.2, the solar image would be 8.2 mm (0.32 in) across.

The set-up for magnified imaging can be very simple: just hand-holding the camera behind the eyepiece can work. Unlike eyepiece projection, in magnified imaging, the separation between the eyepiece and the camera does not affect the size of the image. The separation does, however, affect the field of view seen by the camera through the telescope; the greater the separation, the smaller the field of view. In practice therefore the separation should be kept small. Alternatively the camera can be mounted on a separate tripod (Fig. 4.8), or an adapter made or bought (Fig. 4.9). The separate mounting of the camera and telescope has advantages because shake from the operation of the shutter (and mirror in SLR cameras) is not transmitted to the telescope. The exposures have to be less than the maximum for the effective focal length of the system (Table 4.3) if the Sun's motion is not to produce

---

[46] Binoculars are often specified in this fashion. The first number is the magnification (i.e. ×7 in this case), and the second number is the objective diameter in millimetres (i.e. 50 mm in this case).

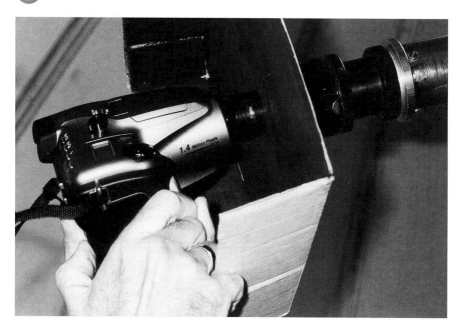

blurring. Thus both the Schmidt–Cassegrain system and the binocular-based system described in the preceding paragraph could be used on an undriven mount and/or with a separately mounted camera since their maximum unblurring exposures with slow film are 1/15 and 1/8 seconds respectively. A directly connecting adaptor also has advantages, being very convenient for a binocular-based system such as that

Figure 4.8.
Magnified imaging with a hand-held digital camera.

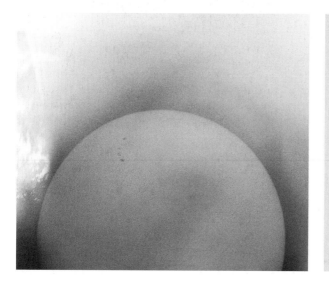

Figure 4.9. Solar image taken through the system shown in Fig. 4.8. Note that the camera lens is larger than the eyepiece and so some stray light is getting in around the edges. This could be avoided by the use of a cardboard tube linking the eyepiece and camera as a shield.

Figure 4.10.
Magnified imaging using binoculars and a welder's #12 filter, with a directly connecting adaptor. Note that the half of the binoculars not being used has been securely blanked off. See Chapter 6 for details of welder's filters. **NB** See the warning about the use of foam for mounting filters with Fig. 2.3.

shown in Fig. 4.10 and producing results like those shown in Fig. 4.11. The unused part of the binocular, however, **must** be securely covered with an opaque screen if the filter does not extend over it. Likewise the separate viewfinder of non-SLR cameras **must** be blanked off.

Figure 4.11. Solar image taken through the system shown in Figure 4.10. A solar image taken at the same time as the images in Figs 4.9 and 4.11 through a 180 mm (7 inch) Maksutov telescope and a metal-on-Mylar full aperture filter is shown in Fig. 1.1, and clearly displays the superiority of the latter system.

Magnified imaging using an SLR camera is simple. It is not critical that the telescope is focused to infinity, since the actual solar image is seen through the viewfinder. It can thus be focused by adjusting the telescope and/or camera while looking through the viewfinder. The image can also be positioned easily where it is required within the frame.

With a non-SLR camera, however, it is critical that the telescope is focused at infinity since the viewfinder will not show the image – indeed the viewfinder must be blanked off to avoid any risk of accidentally glimpsing the unfiltered Sun through it. The apparently obvious way of focusing the telescope is to remove the solar filter and to point it at a terrestrial object at least 1000 metres (1200 yards) away and adjust it to be in focus (but see the discussion in Chapter 7 on the inaccuracies involved in this). Alternatively the telescope can be focused on a night-time object in the sky, and then left set up, or the focus position noted accurately. If you wear glasses for distance vision, then you will need to have them on when focusing the telescope. This method can work, but it is not very precise because of the natural focus adjustment of the eye. In other words, when the telescope is focused near to, but not at, infinity, the eye will automatically adjust itself to bring the image into focus, leaving an incorrect setting for the telescope. Since the adjustment range of the eye decreases with age, this is less of a problem for older observers. If you have focused the telescope in this way then you will need to take several images with a range of focus settings centred around the "infinity" setting in order to be sure of having at least one actually in focus. By keeping a note of the settings on the first occasion that you attempt magnified imaging, the best one can be recorded for future use, so reducing the range of settings needed thereafter.

A more accurate infinity setting can be found from the start if you can borrow an SLR camera or have a second small telescope or pair of binoculars. With the borrowed SLR camera, set the camera focus to infinity and then focus the image in the viewfinder by adjusting **only** the telescope. This can be done on the Sun with the full aperture filter in place or on a night-time or distant terrestrial object if the filter is removed. Once in focus, the telescope setting should be noted, or the instrument left set up. With a small telescope or pair of binoculars, first focus those on infinity. Then, without adjusting the small telescope or binocular setting, use them to look through the main telescope eyepiece (if using the Sun for this, then the main telescope must have a full aperture filter in place), and adjust the main telescope **only** to bring the image into focus. Again if you wear glasses for distance vision you will need to have them on during this process.

Conventional Cameras

A conventional camera and lens can be used to image the Sun, provided that a full aperture solar filter is placed over the front of the lens. Unless the focal length of the lens is half a metre or more the solar image will not be large enough to cover the exposure meter detector of the camera, and an incorrect exposure time will be set. The correct exposure may be found by pointing the camera without the filter in place at a sheet of white card placed perpendicular to the Sun.[47] The required exposure for the Sun through a double layer metal-on-plastic solar filter will be about twice (one photographic stop) that given using the white card. However since filters vary, it is advisable to experiment before using this approach for critical observations (say during a solar eclipse), and to use a wide range of exposures.

CAUTION
REQUIRED

NB If the camera is not a single lens reflex (SLR) design and so has a separate lens for the viewfinder, the filter **must** cover both the imaging lens **and** the viewfinder lens. Care also needs to be taken when sighting through the viewfinder that the eye not being used is not exposed to direct solar radiation. It is strongly recommended that a pair of glasses with the lens for the eye not looking through the viewfinder securely covered with an opaque screen is worn. If you do not normally need to wear glasses, then a cheap pair of non-prescription sunglasses can be used, with the lens for the eye looking through the viewfinder of the camera removed, and the other lens blanked off.

---

[47] The surface brightness of the Sun is about 46,000 times the surface brightness of a piece of white card which returns 100% of the light falling onto it. Most actual samples of white card will reflect 85% or 90% of the light falling onto them, making the Sun brighter by about ×51,000 to ×54,000. A density 5 filter reduces intensities by a factor of ×100,000 (Table 2.1), so when in use on the camera it reduces the direct surface brightness of the Sun to about half that of the card.

Imaging the Sun using a welder's filter (Chapter 6) as a full aperture filter is not recommended on the whole since the image quality is unlikely to be good enough for it to be worth the effort involved. It may, however, be of use when high resolution is not needed; for example, when following the passage of the Moon across the Sun during an eclipse (Chapter 7), and using a conventional SLR camera to obtain the images, a focal length of at least 100 mm will be needed to give even a moderately sized solar image on a 35 mm negative (Table 4.2). A filter with a shade number of 12 or higher must be fixed securely across the front of the lens of the camera. For a non-SLR camera, the filter must cover the viewfinder as well as the camera lens.

CCD Cameras

Most video cameras nowadays are of course based upon CCD detectors, as are digital still cameras. Here therefore we are concerned with the specialist CCD cameras that are sold for night-time astronomical imaging (Fig. 4.12). These usually produce monochromatic images, and therefore (unlike colour film, digital or video cameras) require three separate exposures through different colour filters in order to produce a colour image. However, if the CCD camera can be

Figure 4.12. CCD cameras for night-time astronomical use.

provided with an ordinary camera lens, then in theory all the above photographic methods of imaging the Sun can also be undertaken with the CCD camera.

Paradoxically to astronomers who are usually struggling to detect faint objects, CCD cameras have problems in solar imaging because they are far too sensitive. Typically a CCD detector is 40 or 50 times more sensitive than even fast photographic emulsion. If one were used on an f10 telescope with a full aperture metal-on-plastic filter, the required exposure time would thus be around 0.0005 s (Table 4.1). Since the exposure is actually done electronically within a CCD camera, there is no intrinsic reason why it should not have such a small value. However, in practice few, if any, of the CCD cameras intended for night-time observation, where exposures of seconds to tens of minutes are normal, will actually have software capable of providing submillisecond exposures.

With a normal full aperture filter, an f-ratio of around f100 is needed to bring a CCD camera exposure up to 0.01s (Table 4.2). This either means a very long effective focal length, and hence a large image, so that only a small portion of the Sun will be imaged, or stopping down to very small apertures and so reducing resolution. It may therefore be necessary to double up on filters, or perhaps to use a Herschel wedge (Chapter 2) **as well as** the full aperture filter, in order to reduce the solar intensity to levels that the CCD can manage.

CCD cameras do have one distinct advantage over photographic cameras for solar imaging, and that is when used without a band pass filter, they are most sensitive at wavelengths just into the infrared part of the spectrum. Since the amount of sunlight scattered by the Earth's atmosphere decreases as the wavelength increases,[48] there is proportionally less background light on a CCD image than on a photographic image. The CCD image therefore has a better contrast, and surface features such as granulation are much more easily seen. CCD images can also have their details enhanced by post-exposure image processing (as discussed in the following section).

---

[48] This is why the clear sky is blue: because more blue (short wavelength) solar light is scattered by the atmosphere than red (long wavelength) light.

Image Processing

Introduction

If you obtain images using an astronomical CCD camera or a digital camera, then they can often be improved considerably by processing the images on a computer afterwards.[49] Indeed without such image processing your images may be very disappointing, or even fail to show what you can easily see with the eye. It is also possible to scan photographic images into a computer, or in some cases for the processing firm to produce digital images from photographic negatives. Image processing can then be used to enhance features on photographic images. The same basic processes can be used as for CCD images, but much less improvement will be achieved. This is because the dynamic range of a photograph is much less than that of a CCD image, and so there is less information in the photographic image to be enhanced.

There are two main reasons why initial versions of CCD images may not be all that you could wish for. The first is that a computer monitor can usually display 256 different levels of brightness. The eye can distinguish about 40 levels of brightness, so images on computer monitors are usually adequate. However, the original brightness levels in the image measured by the CCD are to much higher levels of precision than that of either the eye or the computer monitor; typically 65,500 (2^{16}) to 262,000 (2^{18}) levels. Clearly these cannot all be displayed on a monitor with only 256 levels of brightness available. The software associated with the CCD therefore usually chooses a mapping between the CCD levels and the monitor levels. If this were done uniformly, then for a 65,500 level CCD, levels 0 to 255 would all appear as level 0 (i.e. black) on the monitor, CCD levels 256 to 511 would be displayed as level 1 on the monitor, and so on, until CCD levels 65,279 to 65,535 are all displayed as monitor level 255 (i.e. white). While this uniform mapping may not be the best, it will at least show the image's main features. However, some

---

[49] Two image processing packages: Quantum Image and an evaluation version of PaintShop Pro are available on the CD accompanying *Software and Data for Practical Astronomers* by D. Ratledge (Springer Practical Astronomy Series, 1998). Adobe PhotoShop is also widely available and advertised in astronomy magazines as well as in computer magazines.

CCD software will scan through the measured intensity levels in the CCD image, and choose a mapping that displays the majority of the image across the available display levels on the monitor. Thus if you have imaged a sunspot, the sunspot itself may be quite a small part of the whole image. The bright solar photosphere will then dominate the display, and the fainter details of the sunspot will all be assigned to monitor level 0; losing many of the details of the spot. Similarly with night-time images, the majority of the pixels may be very dark and the display will be dominated by them, not the small area covered by the planet, galaxy, etc., that is actually of interest. Details of the sunspot, planet, galaxy, etc., may be brought back through a procedure known as grey-scaling (discussed later).

The second reason why CCD images may be disappointing is that our eyes respond to different intensity levels in a logarithmic fashion. We are also accustomed to photographs, and photographic emulsion responds to light logarithmically as well. A logarithmic response has the effect of stretching the fainter parts of the image and compressing the brighter parts, therefore allowing a wide range of different intensities to be seen or displayed on a photograph. A CCD response, however, is linear and so faint parts of the image appear dimmer, luminous parts brighter, than on the images that we are "used to" (try comparing a photograph of a star cluster with a CCD image of the same object and you will see quite a difference in appearance). In the image of the Sun shown in Fig. 4.13a, the edge of the Sun appears to be running down the left hand side of the image. This, however, is not the true edge of the Sun, but a result of limb darkening and possibly some shadowing or vignetting within the telescope, combined with the linear response of the CCD. Figure 4.13c shows the true edge of the Sun.

Image processing also includes dark noise subtraction, which is probably done automatically by your CCD software. Dark noise is the signal generated within the CCD even when it is not exposed to anything, and mainly arises from thermal motions of the electrons within the silicon substrate of the device.[50] The CCD software will normally take an exposure of the same length as for the real image, but without opening the camera shutter, and then subtract this from the final image.

---

[50] This is why CCDs for astronomical use are usually cooled, so that thermal noise is reduced.

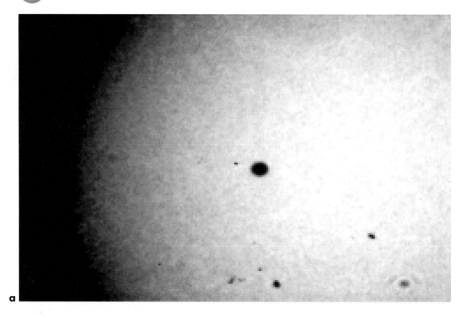

a

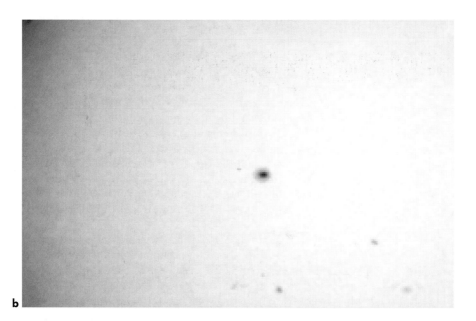

b

Figure 4.13. CCD images of the Sun: **a** towards the centre of the solar disk with an illusory edge resulting from limb darkening and shadowing; **b** the same image grey-scaled to show that the photosphere does cover the whole field of view; **c** the true edge of the Sun.

c

Figure 4.13c.

Other aspects of image processing include true and false colour images, image addition, sharpening, edge enhancement, flat fielding, correction for cosmic rays and bad pixels, and noise reduction. Some or all of these may be available within the software which comes with the CCD camera. Alternatively specialist image processing packages (see footnote 49) are now available widely and quite cheaply.

Photographic images which have been scanned into the computer or produced by a digital or video camera can also be processed. However, such images will not have the dynamic range[51] of a "proper" CCD image, and so the improvements to be made will be more limited. However, operations such as stretching to remove geometrical distortion, zooming, adjusting the colour balance to correct for colour introduced by a filter, or the use of false colour to enhance specific features, can still be useful on such images.

Much of image processing is "trial and error"; you keep on trying different settings and types of processing until you get the image that you want. But it will save time if you understand what the programme is doing to your image. Image processing also divides into data

---

[51] The dynamic range is the ratio of the intensity of the brightest part of the image to that of the faintest. For a photograph it may be up to ×1,000, but for a CCD it may be ×100,000 or more.

reduction and image enhancement. The former is concerned with eliminating or reducing known problems with the image and producing it in a standard form. Data reduction is usually fairly straightforward and clear cut with little judgement being involved. Image enhancement is the production of an optimum image for the purpose that you require. It involves a lot of judgement, and may require the use of several different image processing techniques. The same image may also appear quite differently if enhanced for two different purposes. For example, a CCD image of a solar eclipse would be processed very differently in order to show the outer parts of the corona, compared with revealing details of prominences.

NB It is very easy to ruin an image using image processing. Always make sure therefore that you save the original image separately (preferably on another disk) from the images that you are processing. Work only on copies of the original and you may then always go back to the start if you make a mistake. Also keep a note of the processes that you use and their parameters, so that when you do process an image successfully, you can do it again with the next image.

Data Reduction

If you are obtaining images for display purposes, then you can probably ignore most of this section. However if you are intending to make measurements of position, size, brightness, etc., on your images, then you will need some of the following techniques.

Dark noise subtraction is probably the single most important process, and this has been mentioned above. Almost certainly it will be an option within your CCD control software, or perhaps be done automatically. It is less important for solar work where the image is bright, and the exposures short, compared with the dark background and long exposures when you image night-time objects, but it should still be used. The CCD should also be cooled, if this is possible, in order to reduce the thermal noise further.

If you intend to measure the brightness of points within your image, perhaps in order to determine temperatures or to estimate limb darkening, then flat fielding will probably be needed. The individual detector elements (pixels) within the CCD may vary slightly in their efficiency. The manufacturer of the CCD will normally have corrected this electronically, but the correction is never perfect. The image of a uniformly bright surface will thus not appear uniform, but will be brighter where there is a more efficient than average pixel, and fainter where there is a less efficient than average pixel. By obtaining an image of a uniformly bright surface (a flat field), the actual image may be corrected for the differing responses amongst the pixels. Where the flat field shows a pixel to be (say) 2% more efficient than average, the output from the same pixel within the true image should be divided by 1.02, and so on. Flat fielding is normally an option within the CCD support software. The main problem with the technique is to find a uniformly illuminated surface in order to obtain the flat field image. If your telescope is inside an observatory, then a piece of thick matt white cardboard may be attached to the inside of the dome. With the dome closed, if this cardboard is then illuminated perpendicularly by a single light source as far away as possible from the card, then it may be imaged to provide the flat field (be careful, however, that the telescope itself does not cast shadows onto the card). Alternatively a uniformly grey cloudy sky can be observed during the daytime, or the clear sky during early dusk. Normally you will only need to obtain the flat field image once, though for really precise work, the flat field image should have the same or very similar exposure length and about the same pixel intensities as those of the true image. You may also wish to obtain new flat field images at two or three month intervals in case there are any slow changes in the response of your CCD.

Noise is always present within images even after dark noise subtraction and flat fielding. Its presence leads to loss of detail, poor image quality and even to false features appearing in the image. There are numerous additional techniques for reducing the effect of noise, although it can never be completely eliminated. Providing that the object being imaged is not changing rapidly, then the simplest way of reducing the effects of noise is to make the exposure as long as possible. The exposure length will normally be limited when the brightest parts of the image become saturated (i.e. reaching an intensity of 65,535 for many CCDs). Once this level has been arrived at, further noise reduction can be obtained by adding together several images. Most image processing software has packages enabling images to be added together; they must, however, be in exact alignment when added or detail will be lost rather than gained. This may be a problem if your telescope does not track the Sun accurately.[52] Some image processing packages though will allow you to look at the images to be added together and identify on each the pixels to be mutually aligned (such as the pixel at the centre of a small sunspot, etc.). Noise is reduced when adding images as the square root of the number of images involved; if four images are added, the noise will be halved, if nine images are added, it will be reduced to a third and so on. Another process which reduces noise, though usually at the cost also of reducing the resolution in the image, is smoothing. In smoothing a new image is produced with each pixel in the new image having the average intensity of the same pixel in the old image and of its eight surrounding pixels (Fig. 4.18 below). A smoothed image may appear better looking, but it will normally no longer be suitable for measuring intensities.

One noise problem affects solar images far more than normal astronomical images, and that is the "filling-in" of the darker parts of the image. An image of a sunspot is of a small dark region with very much brighter surroundings. This is the reverse of a normal astronomical image where there is a small bright region (star, planet, galaxy, etc.) with dark surroundings. With the solar image some of the bright light from the surroundings gets scattered into the dark region, making it appear much brighter than it should be. Thus the actual brightness of the centre of a sunspot is about a quarter that of the surrounding photosphere, but it would be unusual for a small telescope at a low-level site to show it as less than half the photospheric brightness. The "filling in" can be reduced by making sure all the optical surfaces of your telescope are clean and dust-free, that mirrors are freshly aluminised, and

---

[52] Remember that the Sun moves relative to the stars, so your normal tracking speed will be incorrect. The normal tracking rate needs to be slowed by 0.27% in order to follow the Sun. However the Sun also moves in declination by about a second of arc every 90 seconds of time on average (Chapter 7). Some computer controlled telescopes are able to track both these solar motions, but most telescopes will not do so.

The unblurred image of a star

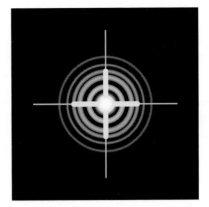

The observed image with the blurring effect of the point spread function

Figure 4.14. A schematic point spread function.

all internal surfaces are painted a matt black.[53] A more complete discussion of ways of optimising your telescope for solar work is given in Chapter 2.

Advanced image processing techniques are also available to tackle the problem of in-filling. These rely upon determining the effect of the telescope and camera upon a point source. This effect is called the instrumental profile, point spread function or PSF (Fig. 4.14, see also Chapter 2). The PSF is used to reduce the blurring in the image either through an iterative procedure known as CLEAN, or in a more sophisticated manner using Fourier transforms. However the details of these procedures are beyond the scope of this book, and the reader is referred to specialist books on image processing for their details and to the help sections of the image processing packages. The PSF may be found from a night-time image, since stars are point sources, the profile of an unsaturated stellar image forms the PSF.

Other more general noise reduction techniques, such as the use of the optimal or Wiener filter, also rely upon Fourier transforms and again are outside the scope of this book. Some image processing packages may

---

[53] This will normally be the case for professionally produced instruments. If it is not, then you should return the instrument to the supplier or manufacturer to have it painted, doing it yourself will certainly invalidate any guarantee, and could result in damage to the instrument.

include regularisation or maximum entropy processing. These are not specifically noise reduction packages but are applied after the noise in the image has been reduced as far as possible. The remaining noise, which can never be eliminated entirely, will leave ambiguities in the image (is that little dot really a small sunspot or is it random noise?). Maximum entropy and related methods enable the "best" image to be chosen from amongst the range of images which fit the data to within the noise levels.

Geometrical distortion of the image may occur for several reasons. For example the image may have been projected onto a screen and imaged from one side, or the shape of the pixels in the CCD may be different from the shapes displayed on the computer monitor. Most image processing packages have a facility for stretching or compressing the image vertically or horizontally, and this can be used to correct the distortion. The degree of stretching or compression required may be determined by imaging an object of known shape (e.g. the whole Sun, which is circular), and adjusting the image until it is seen correctly on the monitor.[54] The same correction factor may then be used for images where the shape is not known in advance.

Image Enhancement

Image processing may be used to enhance images. This is a procedure whereby the image may be altered in various ways with the objective of providing an optimum display of those aspects of the image which are of interest. For example, an image of the Sun may encompass sunspots, granulation and faculae. If the sunspots are the items of interest, then the image may be manipulated to display them to their best effect. This will almost certainly lead to the granulation and faculae being seen less well, or even to them disappearing. However, the original image may then separately be enhanced to show these features, and then the sunspot details will probably be lost. Unlike data reduction, image enhancement is more of an art than a science.

---

[54] A different factor may be needed on printed versions of images, whose value may be determined in a similar manner.

There are numerous processing effects that can be used, and finding the ones which work best is often a case of trial and error. You may also find that enhancing an image for the best display on the monitor is different from enhancing it to get the best printed image. One aspect of image enhancement, however, is fairly straightforward and that is correcting for the effects of cosmic rays and bad pixels.

If a cosmic ray passes through your CCD during the exposure, then it will produce one or two pixels within the image which are much brighter than they should be. When imaging the bright surface of the Sun, it may not be possible to identify such cosmic ray strikes, but they are likely to show up during eclipse imaging. It may also be that your CCD has one or two "rogue" pixels, which are always very bright or very dark, and sometimes such pixels can affect all the others in the same column resulting in a dark streak across the image. Both these effects can be compensated, producing an image with a much more pleasing appearance, although without restoring the original information that has been lost. The compensation is done by taking the average intensity of the eight pixels surrounding the affected one, and using this average value to replace the intensity of the central pixel (Fig. 4.15).

The most useful operation for the majority of astronomical images, including those of the Sun is grey-scaling.[55] Surprisingly some widely available image processing packages do not include this operation, however it may well be a part of your CCD software. Grey-scaling scales the mapping of CCD intensities for the items that you are interested in to the available display levels on the monitor. In practice grey-scaling is undertaken by changing the CCD-to-monitor mapping while looking at the image, until you get the optimum display. However, understanding the background to the process will help to save time. If you look at a histogram of the numbers of pixels with a given brightness (an option in most image processing packages), then a solar image might appear as in Fig. 4.16. Grey-scaling to show the sunspots to best effect would then mean setting intensity level 4000 and

---

[55] In some image processing packages this term may be used for the conversion of a colour image to a monochrome image. Other terms for the process may also be encountered, such as level or threshold adjustment.

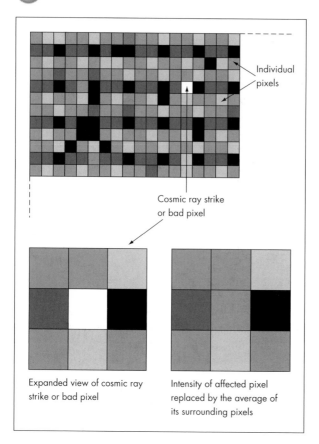

Individual pixels

Cosmic ray strike or bad pixel

Expanded view of cosmic ray strike or bad pixel

Intensity of affected pixel replaced by the average of its surrounding pixels

Figure 4.15.
Compensating for a pixel in a CCD image which has been affected by a cosmic ray strike, or which has a "rogue" response.

below to black (monitor level 0), and intensity level 11,000 to white (monitor level 255), thus with the levels covering the sunspots; 4001 to 10,999, evenly distributed amongst the monitor levels 1 to 254. To show the faculae, would mean distributing the monitor levels 1 to 254 from 19,500 to 21,000, with everything below that range black, and above it white. Since histograms of images may not be as clear cut as the schematic example shown in Fig. 4.16, an alternative is to look at the intensities of individual pixels within the sunspot and within the surrounding photosphere. Most image processing packages will provide a cross-hair on the image which may be moved around using the mouse, and provide a reading of the screen of the intensity of the pixel identified by the cross-hair. An example of grey-scaling and colouring an image is shown in Fig 4.17.

The process of smoothing to reduce the level of noise in an image has already been mentioned. In it the pixel

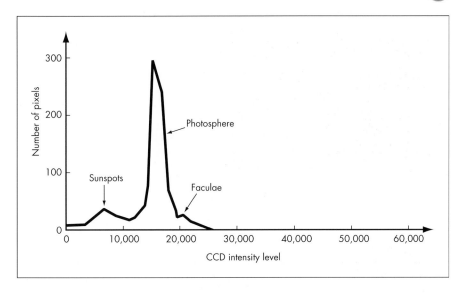

Figure 4.16. A schematic histogram of a solar CCD image.

value is replaced by the mean value of itself and the surrounding eight pixels (or 24 or 48 pixels, etc., for wider smoothing). This is shown visually in Figs. 4.18a, 4.18b and 4.18c, where the numbers in each cell are the multipliers of the intensity in that cell. The new image is obtained by combining the contributions from each of the pixels, weighted by their multipliers, to produce a new value for the central pixel. The results of smoothing are shown in Fig. 4.19c. The opposite process to smoothing is called sharpening.[56] Not unexpectedly, sharpening tends to increase the noise level, but this may sometimes be worth accepting as the price for more clearly seen fine detail. Sample filters for sharpening an image are shown in Figs. 4.18d and 4.18e, and their effects in Fig. 4.19d. A special case of a sharpening filter is shown in Fig. 4.18f, which has the effect of enhancing regions within the image where the intensity is changing rapidly. Since such regions are often edges, the effect of the filter is therefore known as edge enhancement. Its effects are shown in Fig. 4.19e.

Many images have a wide dynamic range. As a result it may be impossible to show the details of both light and dark features simultaneously (Figs. 4.17b and 4.17c, for example). Unsharp masking is a means of reducing the dynamic range and it may allow all parts of the

---

[56] Smoothing and sharpening may be called low pass filtering and high pass filtering in some image processing packages.

a

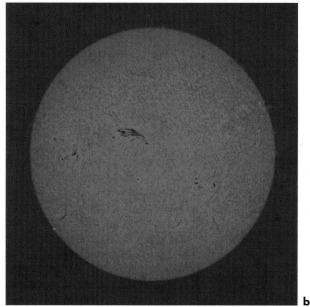

b

Figure 4.17.
The effects of grey-scaling using the Corel PhotoPaint 8 and PaintShop Pro 6 image processing packages: **a** the original monochromatic image of the Sun – a spectroheliogram in Hα light (reproduced by courtesy of Thierry Legault); **b** coloured using PaintShop Pro 6 to represent the actual appearance of the image; **c** grey-scaled using Corel PhotoPaint 8 to reveal the prominences; **d** grey-scaled using Corel PhotoPaint 8 to reveal the main active region.

image to be seen at once (Fig. 4.19f). The operation originated in photography. In photographic unsharp masking a slightly out-of-focus (unsharp) print is made onto film from the original negative. The original negative and its positive are then superimposed and a new print made through the combined films. Where the

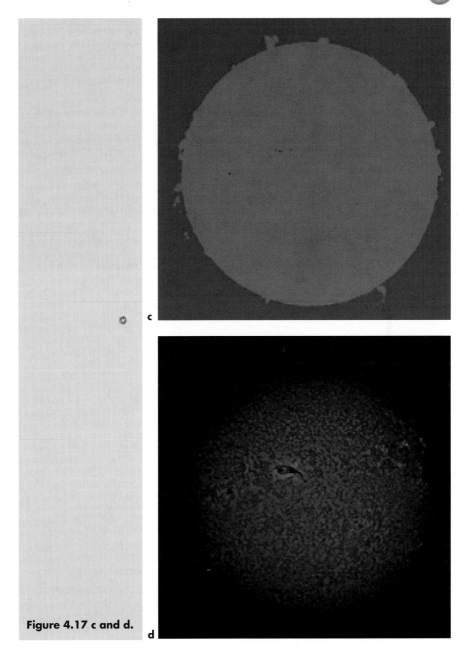

c

Figure 4.17 c and d.

d

negative is darkest, the positive is lightest, and vice
versa. If a sharp positive had been made, then the light
and dark portions of the two films would match exactly
and completely cancel each other out, so that there
would be no resulting image. By using an unsharp

+1/9	+1/9	+1/9
+1/9	+1/9	+1/9
+1/9	+1/9	+1/9

a A smoothing filter

+1/16	+1/16	+1/16
+1/16	+1/2	+1/16
+1/16	+1/16	+1/16

b A weaker smoothing filter

+1/25	+1/25	+1/25	+1/25	+1/25
+1/25	+1/25	+1/25	+1/25	+1/25
+1/25	+1/25	+1/25	+1/25	+1/25
+1/25	+1/25	+1/25	+1/25	+1/25
+1/25	+1/25	+1/25	+1/25	+1/25

c A wider smoothing filter

−1	−1	−1
−1	+9	−1
−1	−1	−1

d A sharpening filter

−0.5	−0.5	−0.5
−0.5	+5	−0.5
−0.5	−0.5	−0.5

e A weaker sharpening filter

−1	−1	−1
−1	+8	−1
−1	−1	−1

f An edge enhancement filter

positive, however, the fine details in the negative do not have corresponding fine detail in the positive, and so the resulting image contains the fine detail of the original, but with the overall dynamic range much reduced. For a CCD image the process is to produce a smoothed version of the original, and then subtract (the equivalent of making a positive print) it, or a fractional multiple of it, from the original.

Other methods of dealing with a wide dynamic range include false colour and contouring. The term false colour is also used to denote a colour image where the original colour is not the same as the displayed colour (see below). Here it is used for a monochromatic image in which the different intensity levels are displayed as different colours. Contouring is also applied to a monochromatic image. The available monitor display

Figure 4.18. Image processing filters: **a** smoothing; **b** weaker smoothing; **c** broader smoothing; **d** sharpening; **e** weaker sharpening; **f** edge enhancement.

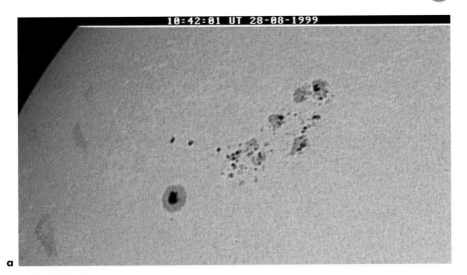

10:42:01 UT 28-08-1999

a

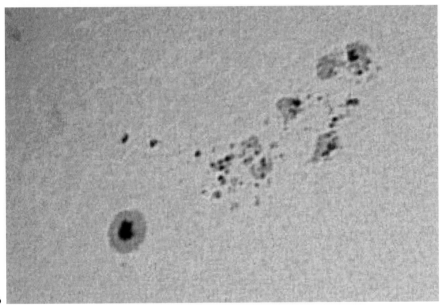

b

Figure 4.19. The effects of filters on an image using PaintShop Pro 6: **a** original white light image of part of the Sun (reproduced by courtesy of Peter Garbett); **b** zoom in to the area of interest; **c** a smoothed image; **d** a sharpened image; **e** edge enhancement; **f** unsharp masking.
(Figure 4.19 c,d,e,f on the following pages)

levels are used repeatedly as the intensity increases. For example in an image with CCD levels from 0 to 1011, the monitor levels 0 to 255 would be used for CCD levels 0 to 255, then from 256 to 511, 512 to 767, and

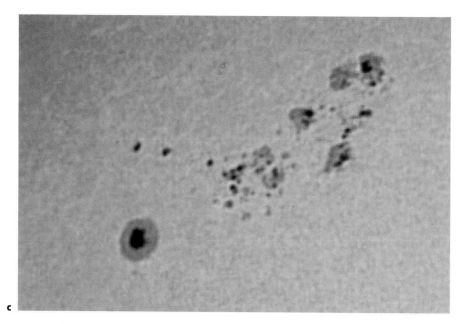

c

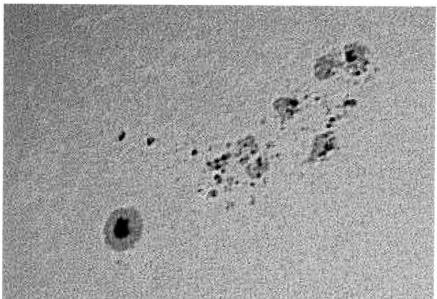

d

finally from 768 to 1011. The resulting images have a series of bands such as that shown in Fig. 4.20. Contouring is most useful for simple images in which the change in intensity occurs in only one direction across the image. It will produce confusing results for complex images.

Figure 4.19 c,d

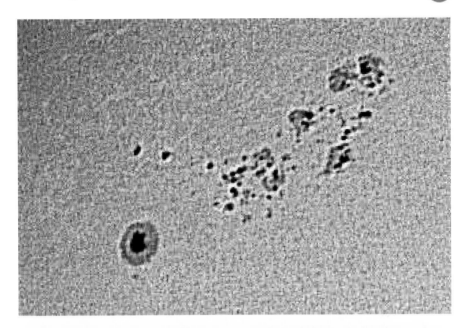

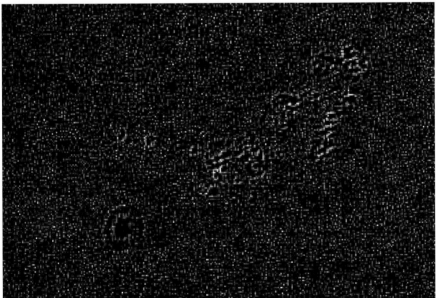

Figure 4.19 e,f

Finally we have true colour imaging and the second type of false colour imaging. A true colour image is obtained either by a colour camera (see the next section), or by a monochromatic camera which takes three images through three different colour filters (usually red, green and blue). In the latter case, the

images are mutually aligned by the image processing package and used to produce a colour image by assigning each original image to the colour of the filter used to obtain it. The levels within the colour image are usually adjustable, so that an accurate rendition of the original colour made be achieved. The same process may be used to correct a colour image obtained through a full aperture solar filter which has left a colour cast (often blue or orange) on the image (Chapter 2). False colour imaging may be used to produce coloured images in which filters have been used to produce monochromatic images in (say) the infrared, red and green parts of the spectrum. These images are then used to produce a displayed image with the infrared original shown in (say) red, the red shown in green, and the green shown as blue.

There are many variants on the image processing operations discussed above, and many operations that have not been considered at all. Very sophisticated processing is possible using Fourier transforms of images, as was seen with the correction of images from the Hubble Space Telescope before its correcting optics were installed. But for these advanced techniques to be attempted successfully a good understanding of the theory of what is happening is needed. Many image processing packages come with useful help sections, but the reader interested in advanced

Figure 4.20.
Contouring on an image of the Sun.

image processing techniques should also consult the specialist literature.

Video Cameras

Video cameras (Camcorders) can be used in place of the photographic, CCD or digital camera in any of the methods just outlined. A full aperture filter will be needed if the camera is pointed directly or through a telescope at the Sun. The S-VHS and Hi-8 formats offer better resolution than VHS or VHS-C, but the resolution on a video image will still be poorer than that of a 35 mm camera image. If the video camera is used with its normal lens and full aperture filter, then the maximum optical zoom, and possibly supplementary lenses, will be needed to obtain a large enough image. Digital magnification can also be used, but since this simply expands the pixel sizes, no additional information is obtained.

Video cameras are ideal for action recording during a solar eclipse either of the eclipse itself, or of the scene around the observing site (Chapter 7).

Chapter 5

Observing Programmes

For many people simply observing the features on the Sun is sufficient in itself. For others, a more long term plan for their observations provides for greater interest. Three possible observing programmes are suggested here. There are many others possible, and anyone with a serious interest in solar work is advised to contact and perhaps join the solar section of their national astronomical society. These often have on-going programmes with many observers contributing so that gaps in the observations due to poor weather or to personal commitments can be reduced.

There is still much to learn about the Sun, and our understanding of the reasons for its phenomena and behaviour is still poor in many areas (Chapter 1). While a lot of work is undertaken by professional astronomers, there is still room for the amateur to make a real contribution, especially if he or she becomes involved with one of the national solar observing programmes.

The Sunspot Cycle

Monitoring the sunspot cycle provides a straightforward, long term observing programme. The sunspot number, R, is determined as discussed in Chapter 1. The new observer will need to find his or her personal correction factor (k) in the formula given in that section. This has to be done by comparing your results with the official values over a period of several months.

The official values are obtainable from the solar section of your national astronomical society or from pages on the internet (search for "sunspot number"). If over a reasonable period of time you find that your values of R are consistently about (say) 90% of the official values, then your value of k is 1.11 (= 100/90). Values of R will fluctuate on a 27-day period because of solar rotation (discussed below), so it is usual to take 27-day averages to eliminate this effect.

The solar cycle is about 11 years long, so keeping a casual eye on the solar activity does not need frequent observations. However, for more serious work, the sunspot number does need to be determined on a daily basis. This is impossible for a single observer, because of interruptions due to the weather or other factors. Monitoring the sunspot cycle is therefore best done as part of a team. While it is possible for several people to get together to set up their own group, most national astronomical societies already have solar monitoring programmes in operation. The new observer can therefore best make a contribution to understanding the Sun by joining such a programme.

Active Regions

Sunspots and sunspot groups start as single small black dots known as pores. These are sunspot umbrae without a surrounding penumbra (Chapter 1), and they may be as small as $1''$ across. The appearance of the pore is preceded by faculae, and when seen through a narrow band filter (Chapters 1 and 8), by plages. In many cases the pore may disappear again after a few hours. In other cases it may grow in size, develop a penumbra, and sometimes be joined by other spots to form a sunspot group.

Large sunspot groups can last for several months. Their lifetime is roughly given in days by one eighth of the spot's maximum corrected area measured in units of a millionth of the area of the solar hemisphere (see Chapter 1). The largest groups can have areas in excess of 1000 millionths of a hemisphere. Thus they can often be observed over several solar rotations, and their development monitored. Associated with a sunspot group will be faculae, flares, prominences, plages (Chapter 8), etc., the whole set of phenomena making up an active region. Often an active region will appear in two halves separated in solar longitude, but with roughly the same solar latitudes. This is termed a bi-polar region. Almost all sunspots are part of a bi-polar region. Even isolated spots usually show a second area marked by faculae and plages where the second part of the bi-polar region exists. Changes within an active region occur on time scales ranging from minutes to weeks; there is therefore plenty of scope for detailed observation.

Unfortunately, solar rotation carries an individual active region behind the Sun once a month for a period of two weeks. Sufficient changes can occur to the active region over that time to make it difficult to identify with certainty when it next reappears, especially if there are several such regions in close proximity. Careful measurement of the region's position on the Sun, and use of the correct rotation period for its latitude (see next section) will be needed to enable re-identification to be made with certainty.

The solar rotation also means that an individual active region is only observed for half the time. Building up a picture of its overall development therefore means observing the development of many regions in order to "fill in the gaps". The appearance of sunspots and groups is classified on the Zurich system[57] into nine classes:

A A single pore, or group of pores, without any bi-polar configuration.

B A group with a bi-polar configuration.

C A bi-polar group; one spot possessing a penumbra.

D A bi-polar group whose main spots have penumbrae. One or more of the spots have some simple structure. The total length of the group is less than $10°$ of solar longitude.

E A large bi-polar group. The two main spots possess penumbrae and complex structures. There are numerous small spots. The total length is generally in excess of $10°$ of solar longitude.

F A very large complex or bi-polar group; length over $15°$ of solar longitude (150,000 km).

G A large bi-polar group without small spots between the main components; length over $10°$ of solar longitude.

---

[57] The recently developed McIntosh scheme has three descriptors. The first letter describes the overall appearance of the sunspot or sunspot group using revised Zurich classes, the second describes the largest spot in a group, and the third, the distribution of the sunspots.

H   A uni-polar spot with a penumbra, and a size greater than 2.5° of solar longitude.

I   A uni-polar spot without a penumbra, and a size less than 2.5° of solar longitude.

Solar Rotation

The Sun is not a solid object and does not rotate like one. The rotational period varies significantly with latitude, by small amounts with time, and may be quite different between the interior and surface. At the equator the rotational period is 24.9 days. It increases away from the equator until near the poles it may be as much as 35 or 40 days. An empirical relationship which gives a good fit for most latitudes is:

Solar rotation period $= 24.9 - 0.0188\ |\phi| + 0.00216\phi^2$ days where ϕ is the solar latitude, north or south, in degrees.

Individual rotations are identified by their Carrington number, based upon a mean solar rotation period of 25.38 days, and starting at midday on 1 January 1854. The start of a Carrington rotation occurs when $0°$ degrees of solar longitude crosses the centre of the solar disk as seen from the Earth. From the Earth the mean solar rotation period, which is also called the synodic solar rotation period, is 27.2753 days, but this varies throughout the year (Table 5.1) because of the varying velocity of the Earth around its orbit. Carrington rotation number 1957 ended and number 1958 started at 14h 40 m UT on 1 January 2000 (Table 5.2). The periods listed in Table 5.1 should be used to interpolate

Table 5.1. Solar synodic rotation period (solar rotation period as seen from Earth) throughout the year based on a true (sidereal) rotation period of 25.38 days

Month	Solar synodic rotation period (days)	Month	Solar synodic rotation period (days)
January	27.309	July	27.242
February	27.301	August	27.249
March	27.287	September	27.264
April	27.269	October	27.281
May	27.254	November	27.298
June	27.244	December	27.308

Year	Carrington rotation number at midnight 1 January (1 Jan. 00)	Carrington rotation number at midnight 1 July (1 July 00)
1990	1824.083	1830.718
1991	1837.465	1844.064
1992	1850.847	1857.519
1993	1864.265	1870.901
1994	1877.647	1884.283
1995	1891.030	1897.629
1996	1904.412	1911.084
1997	1917.831	1924.467
1998	1931.212	1937.849
1999	1944.595	1951.232
2000	1957.978	1964.650
2001	1971.397	1976.931
2002	1984.779	1991.414
2003	1998.161	2004.797
2004	2011.543	2018.215
2005	2024.962	2031.597
2006	2038.344	2044.979
2007	2051.726	2058.361
2008	2065.108	2071.780
2009	2078.527	2085.162
2010	2091.909	2098.544
2011	2105.291	2111.926
2012	2118.673	2125.345
2013	2132.091	2138.727
2014	2145.474	2152.109
2015	2158.856	2165.491

Table 5.2. Carrington rotations from 1990 to 2015

the Carrington rotation number for dates other than those given directly in Table 5.2.

However there are slight variations in the rotational periods obtained, depending upon the methods used to measure them. Thus at a latitude of $45°$, sunspots suggest a rotation period of 28.1 days, while prominences and filaments suggest 27.3 days. This presumably arises through the intrinsic motions of sunspots and/or prominences across the solar surface, and so there is still some uncertainty in the exact rates for the Sun itself.

When sunspots or other features are visible on the Sun it is easy to measure the solar rotation period from their rate of motion across the disk (Figs. 5.1 and 5.2). With two images of the Sun obtained a day or two

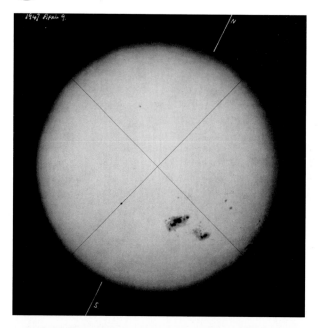

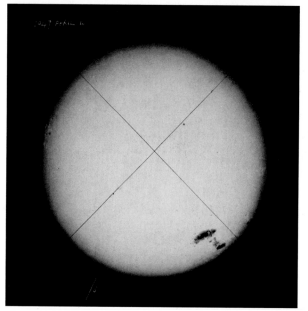

Figure 5.1. A large sunspot group showing the motion across the solar disk due to rotation over a two-day interval (photographs reproduced by permission of the Royal Astronomical Society).

apart, draw a line through the centre of the Sun, parallel to the motion of the sunspot, etc. Then draw a second line through the centre of the Sun at right angles to the first. Measure the distance from the centre line to the solar limb, R, and the two distances D_1 and D_2 (Fig. 5.2). If the two images were obtained at times T_1

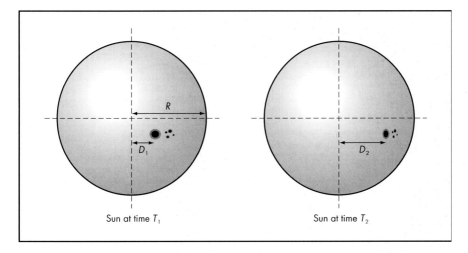

Figure 5.2.
Measurements for
determining solar
rotation.

and T_2, expressed as decimal days, then the solar
rotation period is given by:[58]

Solar rotation rate =

$$\frac{\text{arcsine}\left[\frac{D_2}{R}\right] - \text{arcsine}\left[\frac{D_1}{R}\right]}{T_2 - T_1} \quad \text{degrees per day.}$$

However, our observations are made from the Earth
which is in orbital motion around the Sun. That orbital
motion therefore has to be corrected in order to obtain
the true solar rotation rate. The Earth moves through
$0.986°$ per day around its orbit. The measured rate of
motion of sunspots across the solar disk, in degrees per
day, therefore needs increasing by this figure to give the
true rate, i.e.:

True solar rotation rate =

Measured solar rotation rate $+ 0.986$ degrees per day

The solar rotation period is then given by:

$$\text{Solar rotation period} = \frac{360}{\text{True solar rotation rate}} \quad \text{days.}$$

Since sunspots appear throughout a range of latitudes
during the sunspot cycle (Chapter 1), over a few years

---

[58] The term "arcsine (x)" means "the angle whose sine is x".
The symbol $\sin^{-1}(x)$ is also used for this same quantity. Thus
arcsine (0.3), or $\sin^{-1}(0.3)$ has the value $17.46°$, since sine
$(17.46°) = 0.3$. On calculators it is usually obtained by
pressing the inverse button before the sine button.

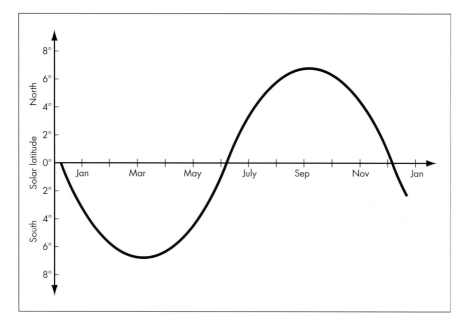

Figure 5.3. The solar latitude of the centre of the visible solar disk throughout the year.

repeated measurements will allow the variation of rotation period with latitude to be confirmed and checked. If you wish to work to high levels of accuracy, then you will need to correct for the inclination of the solar rotation axis to the line of sight. This varies throughout the year. The changing solar latitude of the centre of the visible solar disk during the year is shown in Fig. 5.3. The inclination correction is best accomplished by plotting out diagrams showing the lines of latitude and longitude on the observed solar disk for various inclinations of the solar rotation axis to the line of sight (Fig. 5.4). If these are copied onto transparent overlays at the same size as your solar images (or your images scaled using an image processing package to the same size as Fig. 5.4), then the sunspots' positions can be read off directly. The same plot can be used for inclinations that are positive or negative by inverting the plot for use with the latter. The orientation of the Sun in the plane of the image is shown in Fig. 5.5. The position of north in the sky, and hence the position angle of the north solar pole on the image, may be found by turning off the telescope drive. The image will trail in an east–west[59] direction, and the north direction

Figure 5.4. Solar latitude and longitude lines for inclinations of the solar rotation axis from 0° to ± 7° to the line of sight. Similar grids may be available commercially from telescope suppliers, or from national astronomical societies. They are then sometimes called Porter's solar disks (UK), or Stoneyhurst disks (USA).

▶

---

[59] The western limb of the Sun is sometimes called the preceding limb, and the eastern limb, the following or trailing limb, because of this motion.

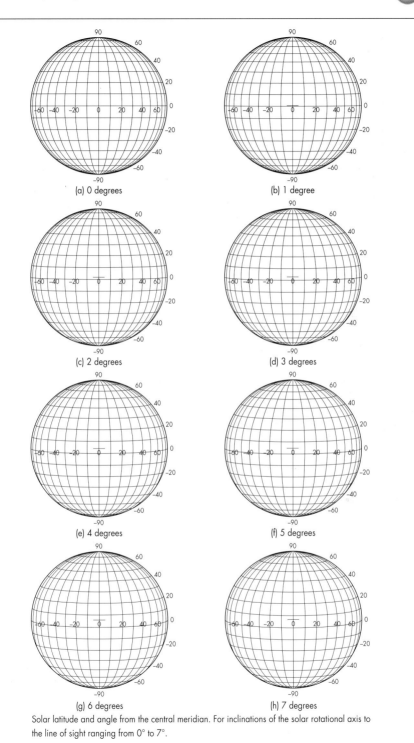

(a) 0 degrees

(b) 1 degree

(c) 2 degrees

(d) 3 degrees

(e) 4 degrees

(f) 5 degrees

(g) 6 degrees

(h) 7 degrees

Solar latitude and angle from the central meridian. For inclinations of the solar rotational axis to the line of sight ranging from 0° to 7°.

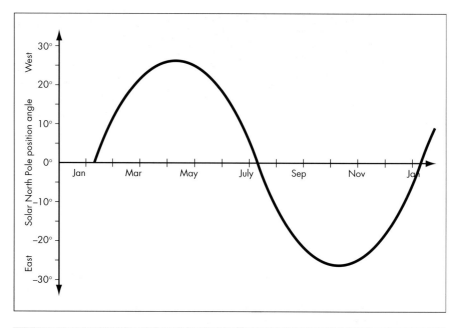

Figure 5.5. The position angle of the north pole of the Sun throughout the year.

in the sky is therefore at right angles to the trail direction. The solar longitude, by convention, is taken to be 0° at the centre of the visible disk at the start of a Carrington rotation.

Chapter 6

Unaided Observations

CAUTION
REQUIRED

In this context unaided observing means observing the Sun at or close to its true angular size of 0.5°. It does **not** mean looking directly at the Sun with the unprotected eye. While it is possible to see the larger sunspot groups without magnification, in practice these types of observations are mostly suited to solar eclipses (Chapter 7). Just as when using telescope or binoculars (Chapter 2) the two safe ways of observing the Sun are either through a solar filter, or by projection.

Projected Images

Figure 6.1.
A pin-hole camera.

When not using a telescope, this means using a pin-hole "camera", but looking at the resulting image and not using photographic film (Fig. 6.1). In other words,

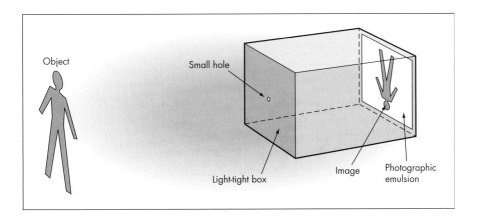

Object

Small hole

Light-tight box

Image

Photographic emulsion

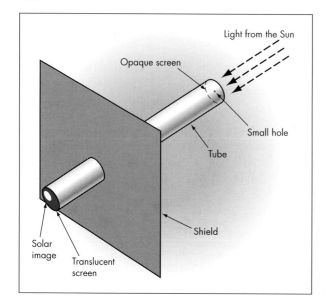

Figure 6.2. A pin-hole viewer for the Sun.

the light from the Sun is allowed through a small hole to fall onto a screen some distance away. Even without the use of a lens or a mirror, if the projection distance is large compared with the size of the hole, then an image of sorts will appear on the screen.

The quality of the resulting image depends upon the size of the hole and upon the projection distance. The size of the image depends upon the projection distance (see Table 4.2 with the projection distance replacing the focal length), and the smallest feature within the image that can be discerned will have to be two or three times the size of the hole. Thus to make out a sunspot group which is 100,000 km long (an angular size of about 2 minutes of arc), the projection distance would need to be about 3000 times the size of the hole, i.e. a projection distance of about 3 metres for a pin-hole 1 mm across (10 feet for a pin-hole 0.04 inches across).

Unfortunately, the use of a tiny hole and a long projection distance, although it is needed to make out fine details of the Sun, means that the resulting image is very dim. In order to see it therefore the image will need to be shielded from the background light from the Sun and from other light sources.

There are innumerable ways of arranging a projected image of the Sun; the possibilities are limited only by your ingenuity. Some of the more widely used options are:

- A large pin-hole viewer. Take a long tube, such as that from the centre of a roll of carpet, a plastic drain pipe, or something similar. Cover one end with an opaque screen which has a small hole punched through it, and cover the other end with a translucent screen, such as ground glass, flimsy typing paper, or tracing paper. Add a shield made from stiff cardboard (or similar – Fig. 6.2), and place on a suitable support with the pin-hole end pointing towards the Sun (Fig. 6.3).

NB Do **not** look along the tube to point it at the Sun – remove the shield, look at the tube's shadow on the ground and adjust the tube's position until the shadow is circular; alternatively and more conveniently, the shadow may be visible on the front of the shield (Fig. 6.4). Once aligned, the image of the Sun should be visible on the screen, and the shield can be replaced in order to see it more clearly. Provided that the hole is small ($\simeq$1 mm or 0.04 inches) and the projection distance greater than about 0.5 metre (1.5 feet), it will be safe to look at the image from behind the screen.

CAUTION
REQUIRED

- A hole in a window blind. This is essentially the same as the pin-hole viewer, except that you, the observer, are now inside the instrument. In a room with a window facing the Sun, black out all the windows. Make a small hole in the blackout towards the Sun. Place a screen made from a sheet of white cardboard, or something similar, a metre or two (a few feet) away from the hole so that the beam of light passing through the hole falls onto that screen. The image of the Sun can then be seen on the screen. Do **not** look through the hole at the Sun.

- Projection onto the ceiling. Onto a suitable window ledge of a window facing the Sun or a table, etc., near to it, place a small flat mirror. Cover the mirror with a sheet of paper which has had a small hole cut into it. The light from the Sun reflected from the unobscured portion of the mirror will form an image of the Sun on the ceiling (Fig. 6.5). Drawing the curtains or otherwise blacking out the room may be needed to see the image clearly. A screen of white cardboard held perpendicularly to the reflected beam will give a less distorted image than that on the ceiling.

- Standing beneath a tree (especially useful during eclipses – Chapter 7). The intersecting leaves of some trees leave small gaps, which act as pin-holes and provide numerous images of the Sun amongst the dappled shade on the ground. Essential additional equipment for this observing procedure includes a comfortable chair and a glass or two of wine!

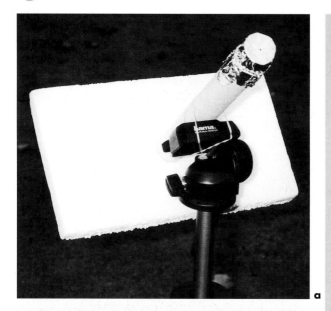

a

b

Figure 6.3. Views of a simple home-made pin-hole viewer (**a**, **b**) and its solar image (**c**).

Filters

Full Aperture Filters

Any filter suitable for use as a full aperture filter (Chapter 2) on a telescope or a pair of binoculars is

c

Figure 6.3c.

CAUTION
REQUIRED

also safe to use for unaided observations. If the filter is physically small, then it should be mounted into a sheet of cardboard, etc., which is large enough to cover both eyes, so that the eye not looking through the filter is shaded from the Sun. Filters made from either metal-on-plastic or metal-on-glass are suitable. However, since both types of filter are highly reflective, it may be helpful to use a screen to shield out back-reflected light from the surroundings. A large cardboard box placed over your head with the filter mounted to cover a hole in one of its sides will work well. Be sure though that the filter is securely attached and cannot fall off while you are looking at the Sun.

The metal-on-plastic solar filters used for full aperture filters are also supplied for unaided viewing, particularly during solar eclipses (Chapter 7), in the form of spectacles. The normal glass lenses are replaced by the filters (Fig. 6.6). Since these devices cover both eyes with solar filters they are safe to use without other precautions. **Do not**, however, be tempted to look at the Sun through a telescope or binoculars while wearing such protective spectacles (see Chapter 2 for safe telescopic methods of observing the Sun).

Figure 6.4. Aligning on the Sun by circularising the shadows – see also Fig. 2.6.

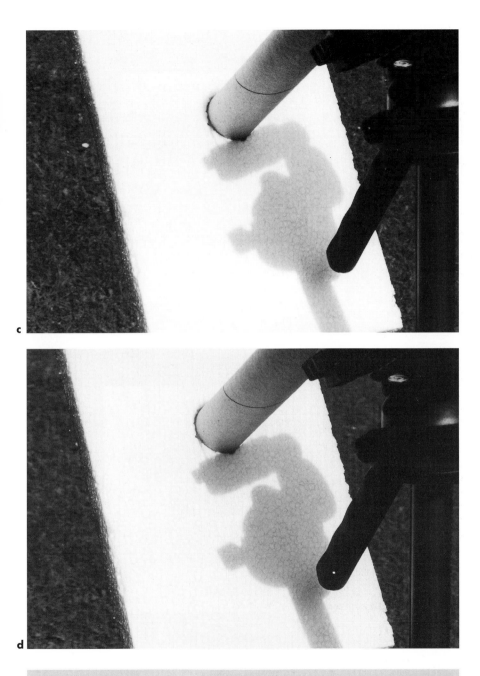

Figure 6.4c,d

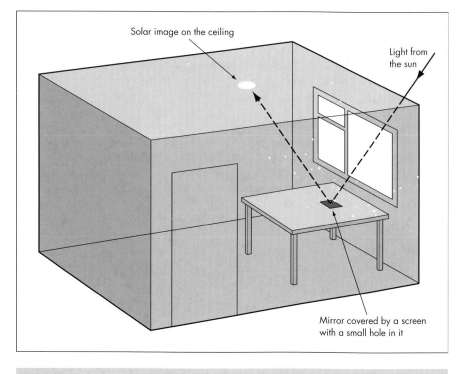

Solar image on the ceiling

Light from the sun

Mirror covered by a screen with a small hole in it

Figure 6.5. A solar image by reflection.

Figure 6.6. Observing the Sun using purpose-made filter spectacles.

Welder's Goggles

CAUTION
REQUIRED

Welder's filters are **not** safe for using at or near the focus of binoculars or a telescope (Chapter 2).

The **only** readily available substitute for a custom-designed solar filter is a filter supplied for use by welders (see the warning at the start of this book). These come in various densities, and shade numbers 12 to 14 or higher (Table 2.1) are needed for safe observing of the Sun. The shade numbers depend upon the visual transmission, and therefore the transmission at other wavelengths may vary depending upon the manufacturer. The ultraviolet transmission of #12 or #14 welder's filters is quite safe for solar viewing since they are designed to protect against the ultraviolet intensity of a welding arc and this is higher than that of the Sun. At long wavelengths though they can be more variable. Thus the transmission curves for two #14 filters produced by different manufacturers are shown in Fig. 6.7. That of manufacturer B has a far higher infrared transmission than that from manufacturer A. In both cases, however, a #12 or #14 filter would be safe for solar work. Although it may be safe, a #12 filter can still give an uncomfortably bright solar image, and so a #14 filter is generally to be preferred.

Welder's filters can be purchased or ordered from large tool suppliers and welding specialists,[60] and are likely to be already mounted into a face mask or goggles (Fig. 6.8). Unmounted filters about 10 cm square (4 inches square) are likely to cost two or three pounds ($5 or so), and four or five times that when ready mounted into a mask or goggles. Filters purchased unmounted should be attached to a suitable cardboard mask so that the whole face is shaded from the Sun. Beware of filters that have actually been used by welders; these may be cracked or scratched or spattered by hot metal droplets from the welding. If undamaged, however, welder's goggles with appropriate shade numbers are safe for viewing the Sun with the unaided eye.

The optical quality of welder's filters is variable, though adequate for unaided viewing. By trying filters from several suppliers, it may be possible to find one

---

[60] See your local business directory for nearby suppliers, if you do not already know of them.

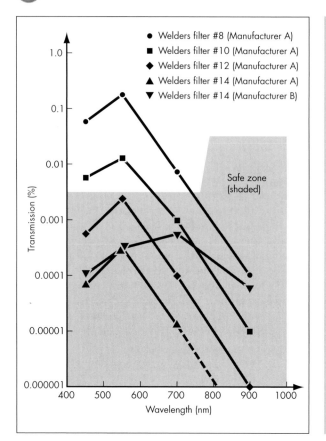

Figure 6.7.
Transmission curves for welder's filters.

Figure 6.8.
Observing the Sun through a welder's mask.

with reasonable optical quality. Such a filter can then be used as a full aperture filter for binoculars. The filter **must** be securely attached in front of one of the objective lenses of the binocular, **and** the unused half of the binocular must be securely covered with an opaque screen. A temporary arrangement suitable for use for a short one-off period is shown in Fig. 2.4. If you intend observing the Sun using binoculars and a welder's filter on a regular basis, then a proper mounting for the filter will be needed. The filters are unlikely to be good enough optically to be used as full aperture filters on telescopes.

Chapter 7

Eclipses

Eclipses

CAUTION
REQUIRED

Warning

Make sure that you have read Chapter 2, and the sections below, before attempting any observations of a solar eclipse of any type or of a transit.

A solar eclipse occurs when the Moon passes between the Earth and the Sun and obscures some or all of the visible solar disk (Fig. 7.1). Eclipses generally are defined as occurring when with two astronomical objects of roughly equal angular size, one object passes across the line of sight to the other. Thus eclipses can also occur between the Galilean satellites of Jupiter and within binary stars. A lunar eclipse is strictly not a true eclipse since it occurs when the Moon passes into the Earth's shadow. Similar alignments can occur when the two bodies are of differing angular sizes, and are called transits when the smaller body is in front of the larger

Figure 7.1. The geometry of a solar eclipse (not to scale).

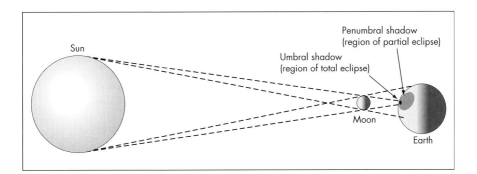

(see the final section of this chapter) and occultations when the larger body passes in front of the smaller.

The physical sizes of the objects involved are not significant in determining whether an eclipse, transit or occultation occurs. The Moon, of course, is physically only 0.25% of the size of the Sun. The Moon's distance from the Earth, however, ranges from 0.24% to 0.27% of the distance of the Earth from the Sun. The two objects therefore have about the same angular size (half a degree) in the sky. This similarity is purely coincidental, and since the effect of tides is to cause the Moon to recede slowly from the Earth, there will come a time when the angular size of the Moon is always smaller than that of the Sun. The present coincidence in angular sizes is, however, useful and fascinating, since during a total solar eclipse the very bright visible disk of the Sun is obscured and the much fainter outer layers – the chromosphere and the corona – become visible.

The Moon's shadow on the Earth is divided into the central umbra and the surrounding penumbra.[61] Within the umbra an observer will see the disk of the Moon completely hiding the normally visible, surface of the Sun (the photosphere). This is a total solar eclipse. The outer layers of the Sun then become visible together with any prominences (Fig. 7.2). Within the penumbral part of the shadow, the solar photosphere is only partly covered by the Moon, and the unobscured segment remains visible as a crescent (Fig. 7.3). This is a partial solar eclipse. If even the smallest part of the photosphere stays visible, then it is so bright that the solar corona will be swamped in its glare. The corona and prominences, etc., thus cannot be seen during partial eclipses.

The Moon's orbit around the Earth is quite elliptical, and so its distance from the Earth varies. The angular size of the Moon in the sky therefore also varies from 29′ 23″ to 34′ 09″. The Earth's orbit around the Sun is also elliptical, though less so than that of the Moon. The angular size of the Sun in the sky thus varies from 31′ 28″ to 32′ 32″. On occasions, therefore, the Moon's angular diameter is less than that of the Sun. If an eclipse occurs at such a moment, then even if the disk of the Moon passes centrally over the Sun, the whole of

---

[61] The inner and outer parts of a sunspot are also called the umbra and penumbra (Chapter 1), by analogy with this structure for a shadow. There is not usually much chance of confusion between the two usages.

Figure 7.2. A total eclipse of the Sun showing prominences and the inner part of the corona (photograph reproduced by permission of the Royal Astronomical Society).

Figure 7.3. A partial eclipse of the Sun with the image being produced by eyepiece projection.

the photosphere will not be obscured and there will not be a total eclipse. Such eclipses, where a thin ring of the photosphere remains visible at all times, are called annular eclipses. Though of some interest as a phenomenon, they do not allow the chromosphere or corona to be seen and so do not cause the same level of excitement amongst observers as total eclipses.

The Earth is rotating and both the Earth and Moon are moving around their orbits.[62] The shadow of the Moon therefore moves rapidly over the surface of the Earth. The maximum diameter[63] of the umbral part of the shadow is about 270 km (170 miles), but the average value is around 160 km (100 miles). Hence for a fixed observer totality lasts for only the short interval while the umbral shadow is passing over his or her position on Earth. The theoretical maximum duration for a total eclipse at the equator is 7 minutes 58 seconds and at latitude 50° north or south, it is 6 minutes 10 seconds. But total eclipses lasting longer than 7 minutes are rare, and the average duration is only 3 or 4 minutes. Since the Moon will be near its greatest distance from the Earth when an annular eclipse occurs, it will be moving relatively slowly. The duration of the annular phase of an eclipse can therefore be up to 12 minutes 24 seconds.

The penumbral part of the Moon's shadow can be up to 3850 kilometres (2400 miles) across at the Earth's surface. The partial phases of the eclipse which precede and follow the total phase can thus last for over an hour each. A partial eclipse which does not become total (i.e. one seen by an observer to one side of the track of the umbral part of the shadow across the Earth's surface) can last for two hours or more. Since the penumbra covers a much larger area of the Earth's surface than the umbra, far more people are in a position to see the partial phases of the solar eclipse than the total phase. To put this another way, at any given spot on the Earth, there will be several tens of partial solar eclipses to be seen for every total eclipse. On average a total eclipse will occur at a specific place about once in 350 years.

A solar eclipse requires the Moon to pass between the Earth and the Sun. If the Moon's orbital plane were

---

[62] The rotational velocity of the Earth varies from zero (at the poles) to 0.47 km/s at the equator. The Earth's orbital velocity is 29.9 km/s and that of the Moon is 1 km/s.

[63] This is the diameter perpendicular to the line to the Sun. The projection of the shadow onto a sloping part of the Earth's surface can make it much larger. This does not, however, increase the eclipse duration since the projection onto the sloping surface will increase the shadow's velocity by a similar value to its increase in size.

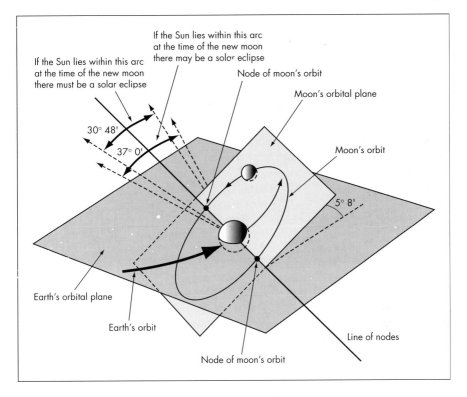

If the Sun lies within this arc
at the time of the new moon
there may be a solar eclipse

If the Sun lies within this arc
at the time of the new moon
there must be a solar eclipse

Node of moon's orbit

Moon's orbital plane

30° 48'

37° 0'

Moon's orbit

5° 8'

Earth's orbital plane

Earth's orbit

Line of nodes

Node of moon's orbit

Figure 7.4. The Moon's orbit and limits for solar eclipses.

parallel to the orbital plane of the Earth around the Sun, then the Moon would produce an eclipse at every new moon, i.e. 12 or 13 eclipses per year. In fact the Moon's orbit is inclined to the Earth's orbit by just over 5°. This angle is sufficient that at most occurrences of the new moon, the Moon's shadow passes above or below the Earth, and there is no visible eclipse.

For a solar eclipse to be seen on Earth, the new moon must occur at a time when the Moon, in its passage around its orbit, is close to the plane of the Earth's orbit (Fig. 7.4). These two points (one on either side of the Earth) are called the nodes of the Moon's orbit (Fig. 7.4). The intersection of the two orbital planes produces the line of the nodes which passes through the nodes and the centre of the Earth. For a solar eclipse of any type to occur on the Earth, the new moon must occur when the line of nodes points to within 18° 30' of the Sun. Because of the ellipticity of the Moon's orbit, a solar eclipse will not always occur even if the line of nodes does point this close to the Sun. However, if the line points to within 15° 24' of the Sun, then a solar eclipse must occur at that new moon. If there is to be a possibility of the centre of the Moon's shadow falling onto the Earth's surface; leading to a total or annular eclipse, then the line of nodes must be within 11° 50′ of the Sun at the time of new moon. To guarantee a total or annular eclipse, however, the line of nodes must be within 9° 55′ of the Sun.

These limits on the occurrence of solar eclipses mean that in any year a minimum of two and a maximum of five solar eclipses will occur. The average number is about 2.3, and of these, about 35% will be entirely partial eclipses, about 35% will be total at least for some of the time, and about 30% will be annular eclipses. For more than two eclipses to occur in a year, they have to take place near the outer limits of the allowed ranges. In such years therefore the eclipses all tend to be partial. Thus there are four eclipses in each of the years 2000 and 2011, but they are all partial (Table 7.1).

Since solar eclipses occur when the Earth in moving around its orbit causes the line of nodes of the Moon's orbit to point to within a few degrees of the direction of the Sun, there are thus two positions on opposite sides of the Earth's orbit when eclipses can occur. But the Earth's position in its orbit corresponds to the months of the year, and so there are two periods in the year when eclipses can occur. At the time of writing, eclipses are occurring in February/March and August/September. If the Moon's orbit were fixed in space then these months would be the only times when eclipses would ever occur. The Moon's orbit, however, is affected by a great many factors: the gravitational fields of the Earth, Sun and planets, tides, the equatorial bulge of the Earth, and so on, so that it does not remain stationary in space.

There are two main changes to the Moon's orbit. Firstly the plane of the Moon's orbit rotates while keeping the same angle to the plane of the Earth's orbit. This causes the line of nodes to rotate, and they complete 360° of movement in 18.61 years. Secondly the perigee[64] of the Moon's orbit moves around its orbit every 8.85 years. The first effect changes the months when solar eclipses occur, while the second causes eclipses to be total or annular accordingly as the perigee or apogee of the orbit is close to the node. Inspection of

---

[64] The perigee is the point in the orbit when the Moon (or any other object orbiting the Earth) is closest to the Earth. The apogee is the point in the orbit when the Moon is furthest from the Earth.

Table 7.1 will show that in 1995 eclipses occurred in April and October. The eclipses then cycle through all the months over the next 18 years:

1998	February and August
2001	December and June
2004	October and April
2007	September and March
2010	July and January

finally returning to April and October in 2014.

Through a quite remarkable series of numerical coincidences, eclipses occur in cycles. The period of the cycle is called the saros and its length is 6585.32 days (18 years 11.32 days[65]). This period is equal to 223 times the lunar month.[66] The interval between successive alignments of the Sun and the same node of the Moon's orbit is called the eclipse year, and it has a value of 346.62 days.[67] Nineteen eclipse years therefore amount to 6585.78 days, remarkably similar to the length of 223 lunations. However there is an even more remarkable coincidence. The perigee of the Moon's orbit is also in motion as we have seen, and so the interval between successive passages of the Moon through perigee (known as the anomalistic month) has a value of 27.554 days. 239 anomalistic months therefore amount to 6585.54 days. Finally in a yet further coincidence that borders on the incredible, the line of nodes will have returned to its previous position after 6585 days as well. Now 6585 days is 18.03 years, which is rather different from the 18.61 years it takes the line of nodes to revolve through 360°. But after one saros interval, the line of nodes will be 11.2° short of returning to its previous position. The Earth, of course, completes its orbit in one year, and in the extra 11.32 days required to complete the saros interval, it will have moved a further 11.16° around its orbit, almost exactly the amount required to make up the lag in the motion of the line of nodes.

---

[65] Sometimes it is 18 years 10.32 days depending upon how leap years fall within the interval.

[66] The interval between successive new moons (or full moons, etc.); its value is 29.531 days.

[67] Not 365.25 days because the motion of the line of nodes means that the Sun needs to move through less than 360° in order to repeat its alignment with the node.

Thus 18 years 11.32 days after one eclipse occurs, the alignment that produced it will be almost exactly duplicated; the Moon will be at the same point in its orbit, the Moon's position with respect to perigee will recur, the Sun will be at the same angular distance from the line of nodes, and the line of nodes will be in the same direction in space. There will therefore be another eclipse, whose properties will be very similar to the one occurring 18 years earlier (i.e. partial, annular or total, and of a similar duration, etc.). There is one difference between successive eclipses in a saros cycle, and this arises from the fraction of day by which the period exceeds a whole number of days. This extra 0.32 days means that when the next eclipse occurs, it will be roughly eight hours later in the day than the earlier one. The Earth will therefore have gone through a third of its rotation and the next eclipse will occur about 120° to the west of the earlier one. However three saros cycles amount to almost exactly 19,756 days (54 years and 34 days). After that interval an eclipse will repeat at or close to the same point on the Earth. The saros cycles are numbered, and the next new cycle, which will be number 156, starts with the partial eclipse on the 1 July 2011.

The slight mismatch between the numbers involved in the above calculations means that there is a slow change to the nature of the eclipses. A particular series of eclipses will start near the Earth's north or south pole, and then move towards the opposite pole by about 300 km (200 miles) for each successive eclipse. Allowing for partial eclipses at the start and finish of the series where the centre line of the shadow misses the Earth, a complete series of eclipses will last for about 1300 years and contain some 70 or so eclipses. At the start of a saros series the eclipses will be partial. They are likely to be annular as the centre line of the shadow first starts to intersect the surface of the Earth, with more of them becoming total as the series proceeds. The pattern will reverse to wards the end of the series. At any given time there are about 40 saros cycles in progress.

The movement of the Moon and the changes in its orbit are very complex, but are now well understood. Its position and that of the Earth can therefore be predicted many centuries into the future, or back into the past. Solar eclipses (and lunar eclipses) can therefore also be predicted well in advance, together with the tracks over the Earth's surface of the umbral and penumbral parts of the Moon's shadow (Fig. 7.5).

Table 7.1. Solar eclipses from 1995 until 2025: data from J. Meeus, C.C. Grosjean and W. Vanderleen, *Canon of Solar Eclipses* (Pergamon Press 1968)

Date	Year	Type[a]	Saros	Land areas from where total/annular phases are visible (if applicable), or the maximum magnitude[b] for partial eclipses
29 April	1995	A	138	Equatorial S. America
24 October	1995	T	143	Southern Asia, S.E. Asia
17 April	1996	P	148	88%
12 October	1996	P	153	76%
9 March	1997	T	120	Arctic, N.E. Siberia
2 September	1997	P	125	90%
26 February	1998	T	130	Panama
22 August	1998	A	135	Indonesia
16 February	1999	A	140	Australia
11 August	1999	T	145	Europe, S. Asia
5 February	2000	P	150	58%
1 July	2000	P	117	48%
31 July	2000	P	155	60%
25 December	2000	P	122	72%
21 June	2001	T	127	Southern Africa
14 December	2001	A	132	Costa Rica
10 June	2002	A	137	Almost entirely over the Pacific Ocean
4 December	2002	T	142	Southern Africa
31 May	2003	A	147	Iceland
23 November	2003	T	152	Antarctica
19 April	2004	P	119	74%
14 October	2004	P	124	93%
8 April	2005	T/A	129	Panama, Colombia, Venezuela (annular except over the Pacific)
3 October	2005	A	134	N.E. Africa
29 March	2006	T	139	Central Asia, N. Africa
22 September	2006	A	144	N.E. South America
19 March	2007	P	149	88%
11 September	2007	P	154	75%
7 February	2008	A	121	Antarctica
1 August	2008	T	126	Greenland, Arctic, Siberia, China
26 January	2009	A	131	Indonesia
22 July	2009	T	136	China, S. Asia
15 January	2010	A	141	Central Africa, S.E. Asia, China
11 July	2010	T	146	Cape Horn
4 January	2011	P	151	86%
1 June	2011	P	118	60%
1 July	2011	P	156[c]	10%
25 November	2011	P	123	91%
20 May	2012	A	128	S.W. USA, Japan, China
13 November	2012	T	133	Australia
10 May	2013	A	138	Australia
3 November	2013	T/A	143	Central Africa (total)
29 April	2014	A	148	Antarctica

Table 7.1. (*continued*)

Date	Year	Type[a]	Saros	Land areas from where total/annular phases are visible (if applicable), or the maximum magnitude[b] for partial eclipses
23 October	2014	P	153	81%
20 March	2015	T	120	Arctic
13 September	2015	P	125	79%
9 March	2016	T	130	Indonesia
1 Sept	2016	A	135	Southern Africa
26 February	2017	A	140	Cape Horn, Southern Africa.
21 August	2017	T	145	Central USA
15 February	2018	P	150	60%
13 July	2018	P	117	34%
11 August	2018	P	155	74%
6 January	2019	P	122	72%
2 July	2019	T	127	Chile, Argentina
26 December	2019	A	132	Arabia, India, Indonesia
21 June	2020	A	137	China, S. Asia, Arabia, Central Africa
14 December	2020	T	142	Chile, Argentina
10 June	2021	A	147	Canada, Arctic, W. Siberia
4 December	2021	T	152	Antarctica
30 April	2022	P	119	64%
25 October	2022	P	124	86%
20 April	2023	T/A	129	Indonesia
14 October	2023	A	134	Central America, Northern South America
8 April	2024	T	139	Mexico, S and E. USA.
2 October	2024	A	144	Cape Horn
29 March	2025	P	149	94%
21 Sept	2025	P	154	85%

[a]T = total eclipse, A = annular eclipse, T/A = eclipse annular for part of the time, and total for part of the time, P = partial eclipse only (i.e. no part of the centre of the Moon's shadow intersects the surface of the Earth)

[b]The maximum magnitude of a partial eclipse is the maximum area of the solar disk obscured by the Moon.

[c]This is the first eclipse in a new saros series.

Sources for these predictions are listed in Appendix 1. The eclipses to occur over the next 20 years or so are listed in Table 7.1.

Eclipse Expeditions

Observing total solar eclipses differs in several ways from observing the Sun generally, but one of the most crucial is that you have to wait for the eclipse to occur

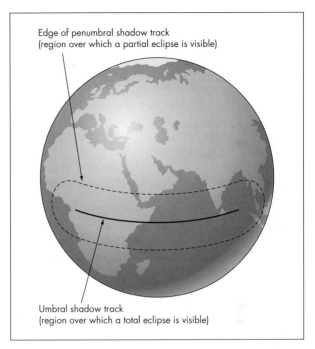

Edge of penumbral shadow track
(region over which a partial eclipse is visible)

Umbral shadow track
(region over which a total eclipse is visible)

Figure 7.5. Track of the Moon's shadow over the Earth (schematic) showing the areas from where total and partial eclipses will be visible.

and to be in the right place when it does. As we have seen, the umbral shadow of the Moon is at most 270 km wide, and is usually nearer to half this value. The movements of the Moon and Earth cause the Moon's shadow to travel over the Earth's surface, so that the total phase of the eclipse will be visible anywhere along the track of the umbral shadow (Fig. 7.5). The length of the umbral shadow track can be several thousand kilometres, so that it might seem to be easy to find somewhere convenient along the track from which to observe the eclipse. In practice a law of perversity seems to operate so that for many total eclipses much of the track of totality lies over the ocean. The schematic track shown in Fig. 7.5 is quite typical; the only land-based observing positions would be in mid-Africa and Sri Lanka.

In addition to the limited geographical areas covered by the total eclipse track, an intending observer must also take into account the likely weather conditions,[68]

---

[68] The eclipse itself is likely to affect weather conditions, since the incoming energy from the Sun is greatly reduced along the eclipse track. Unfortunately there is no consistent pattern to the weather changes brought about by the eclipse; conditions are as likely to get worse as to improve.

the quality of the observing site (especially dust and haze from man-made or natural sources), the altitude of the Sun at the time of the eclipse, the duration of the eclipse (which varies along the path), access and communications to the area, and by no means least, the political stability of the area. When all these factors are taken into account, suitable observing sites for most eclipses are quite restricted, and observers often have to travel considerable distances in order to have a good chance to see the eclipse.

In the past trips to observe total eclipses have sometimes been quite heroic affairs with tons of equipment and numerous personnel having to be transported across seas and deserts in expeditions that lasted for many months. Even today, organising an eclipse expedition is not trivial, since the observing site is unlikely to be near popular tourist destinations with their well organised and inexpensive ready made transport and accommodation. Additionally, if more than the simplest observations are to be made, then telescopes, filters, cameras, power supplies, computers, shelter and storage, etc., may all need to be transported to the site. It is possible for all of this to be undertaken by an individual, but there are many advantages in several people joining together for the expedition and in sharing the burdens.

Local and national astronomical societies often organise trips to relatively nearby eclipses or to those which may have some especial interest. In recent years, however, there have been several travel firms which have started to offer eclipse expeditions on a commercial basis. These are usually advertised well in advance of the eclipse in the popular astronomy magazines, and sometimes in the national press. In addition to having most of the work done for you, on these tours the eclipse is usually only a part of the whole experience. If, as frequently happens, the eclipse is clouded out, you will therefore still have had a holiday in what is probably an exotic and interesting part of the world to look back on. A further advantage of a specialist eclipse holiday is that it can form a part of a cruise trip. Since the total eclipse track often lies over water (Fig. 7.5), this greatly expands the available area for the observations and increases the likelihood of seeing the event, because to a limited extent the ship can move to the best spot given the weather conditions on the day.

On some occasions, trips may be offered on aircraft to observe eclipses. This has two advantages. The first is that in most cases the aircraft will be able to climb above the cloud and so almost certainly guarantee a view of the eclipse. The second advantage is that the aircraft can fly along the eclipse path, and so extend the duration of the total phase of the eclipse. If the aircraft is supersonic, like Concorde, then this effect can double the natural length of totality. The disadvantage of observing from an aircraft is that you will be looking through small windows which usually have considerable distortion, and there will be little point in or opportunity for using a telescope. A few aircraft have been specially adapted with optical windows and gyroscopically stabilised platforms to allow telescopic observation. But these are only available to professional observers, and time on them is allocated, like that on major ground-based telescopes, through competition.

Naked Eye Observing

CAUTION
REQUIRED

Once the eclipse is total, it is quite safe to observe the solar corona with the unprotected eye. The coronal intensity is around 35% to 50% of the intensity of the full moon. **But, so long as any part of the normal surface of the Sun (the photosphere) remains visible, all the precautions for solar observing discussed in Chapter 2 must be followed.**

If you are known to be interested in astronomy then you will invariably be asked about observing eclipses when a nearby one occurs. You may also be involved with a group of people or even leading it, for observing the eclipse. It is important to stress the dangers of looking at the Sun directly. Particularly with young children, there is always a temptation to "try it for themselves", perhaps by squinting through nearly closed fingers. Make sure that everyone always uses a safe observing method (Chapter 2). Projection methods (but note the warning in Chapter 2), or the use of commercially produced "eclipse spectacles" (Fig. 6.6), provided that everyone has a pair, are probably the best approaches when involved with a group.

On the other hand, observing an eclipse is no more dangerous than observing the uneclipsed Sun. Sometimes well-meaning warnings on the dangers of solar observing are given so much prominence at the time of an eclipse that some people may avoid watching it altogether. There are even cases where families have hidden in cellars under the impression that the eclipsed Sun was much more dangerous than the uneclipsed Sun. All phases of an eclipse can be safely observed using the methods given in Chapter 2, and the "total" phase may be quite safely observed with the unprotected eye, through binoculars or through a telescope, although in the latter cases make sure that you stop using the instrument some seconds before the end of totality.

A total eclipse is preceded and followed by partial eclipses as the Moon's disk progressively covers the Sun. The full sequence of events (Fig. 7.6) follows the same pattern on each occasion with minor variations due to the circumstances of the eclipse and changes in the Sun due to the sunspot cycle (Chapter 1). The sequence is:

- The start of the eclipse (first contact) when the Moon's edge is first discernible against the edge of the Sun.
- The partial phase, lasting about an hour during which the Moon occults more and more of the solar disk (Fig. 7.3). Observing this phase of the eclipse, indeed any part of the eclipse wherein the solar photosphere is still visible, may be undertaken by any of the methods discussed in Chapter 2 for observing the uneclipsed Sun. Multiple images of the partial phase may often be seen beneath trees where the gaps between the leaves act as the entrance apertures to pin-hole cameras (Chapter 6).
- Baily's beads: for a few seconds just before totality, the thin crescent of the solar photosphere develops a series of bright knots, very much resembling the beads on a length of string. These occur where there are valleys on the edge of the Moon. The valleys temporarily allow more of the photosphere to be seen giving the impression of a brighter region. The effect is mainly due to irradiance[69] within the eye, as the "beads" actually have the same surface brightness as the remaining crescent of the solar disk. Since even a

---

[69] The eye tends to see small bright objects as larger than reality when they are against a dark background. This is because of cross-connections between the cone and rod cells which are the light receptors in the retina of the eye. The cross-connections mean that cells outside the brightly illuminated part of the retina also respond to the incoming light.

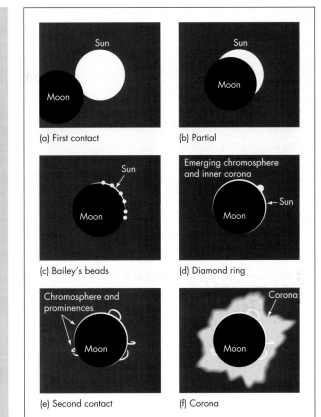

Figure 7.6. The pictorial sequence of events throughout a total eclipse of the Sun (the sequence recurs in reverse after totality).

small change in the attitude[70] of the Moon will completely change the valleys seen along the edge, the positions of the beads vary from one eclipse to another.

---

[70] Most people know that the Moon always keeps the same face towards the Earth. It might therefore seem that the valleys would be the same at every eclipse. However the reason that we always see the same side of the Moon is that, through the effects of tides, its rotational and orbital periods are exactly equal. But the Moon's rotational velocity is constant, whereas its orbital velocity varies because of its elliptical orbit. Sometimes therefore the Moon's rotation "gains" a little on its orbital motion, and at other times it "loses". We can thus on some occasions see parts of the Moon at its edges which would normally be on its far side. This effect is called libration, and through it we can see about 59% of the Moon's surface from the Earth. Libration movement also means that the valleys at the edge of the Moon during solar eclipses change, so also changing Baily's beads.

Figure 7.7. The solar corona during a total solar eclipse (photograph reproduced by permission of the Royal Astronomical Society).

- The diamond ring: the last sliver of the photosphere is still to be seen as the chromosphere and the inner parts of the corona start to become visible.
- The start of the total phase of the eclipse (second contact) occurs as the last portion of the photosphere is occulted by the Moon. Small red features will be seen projecting from the Sun beyond the edge of the Moon. These are the prominences (Fig. 7.2, Chapters 1 and 8).
- The corona becomes more easily seen as the eclipse deepens and your eyes improve their dark adaption. The corona varies with the sunspot cycle (Chapter 1), sometimes being a circular pearly glow around the Sun, on other occasions appearing with long streamers (Fig. 7.7). Smaller scale radial streamers are often seen around the solar polar regions and are called coronal rays.
- The total phase of the eclipse ends with the third contact, when the diamond ring effect reappears with reversed orientation as the trailing limb of the Moon uncovers the first portion of the solar photosphere.
- The opening sequence of the eclipse now appears in reverse, with the reappearance of Baily's beads, another hour-long partial eclipse phase, and finally the complete ending of the eclipse at fourth contact, when the Moon finally uncovers the last fraction of the photosphere.

The solar corona can be well seen with the unaided eye, but its fine details and those of the chromosphere and prominences will need the use of binoculars or a telescope (see the next section). There are, however, other features to look out for without optical aid during totality which do not relate directly to the Sun. Thus it may be possible to see stars. The corona is not quite as bright as the full moon, and so the brighter stars should be visible given reasonable observing conditions. This will enable the position of the Sun in the sky to be noted, and if previous eclipses have been observed, changes in that position will be seen.[71] The changing position of the Sun against the background stars is familiar indirectly through the change in the constellations to be seen at night. But it is only during eclipses that its position can be determined directly. The

---

[71] Contrary to most people's expectations, the Sun in its yearly motion passes through 13 constellations, not the 12 normally associated with the zodiac. The extra constellation is Ophiuchus. The dates when the Sun is within a given constellation get later in the year through the effect of precession, at a rate of about 1 day every 70 years. The irregular boundaries of the constellations mean, however, that this rate of change is only an average value. At the start of the twenty-first century, the dates are:

Constellation	Dates of solar passage
Sagittarius	19 December to 21 January
Capricornus	22 January to 16 February
Aquarius	17 February to 12 March
Pisces	13 March to 18 April
Aries	19 April to 14 May
Taurus	15 May to 21 June
Gemini	22 June to 21 July
Cancer	22 July to 11 August
Leo	12 August to 17 September
Virgo	18 September to 31 October
Libra	1 November to 22 November
Scorpius	23 November to 30 November
Ophiuchus	1 December to 18 December

These times of actual solar passage are quite different from the fictitious usage of astrologers. Thus the Sun is in the Zodiacal sign of Sagittarius from 22 November to 21 December, which only overlaps with reality for two days. The true times taken for the Sun to pass through a constellation are also quite variable: from 8 days in Scorpius to 44 days in Virgo, rather than the uniform month assumed for each constellation by astrologers.

brighter planets, if above the horizon, should be visible, and this is a good opportunity to look for Mercury which will be less than 27° from the Sun.

Additionally there are the terrestrial effects of a total eclipse. The reduction in light intensity can fool animals, birds and plants into assuming night fall. Animals and birds may therefore go to sleep or start roosting and flowers may close up. They come back to "life" though with startling rapidity at the end of the eclipse. Also, as mentioned above, the local weather may change through the reduction in solar energy arising from the eclipse.

The other phenomena which can sometimes be seen for a minute or two preceding or following totality are the shadow bands. These are observed as slow moving, low contrast bands of light and dark on white or pale coloured surfaces. Shadow bands are quite variable; sometimes they are not seen at all, sometimes they may be stationary, at other times they move at a few metres per second (10 feet per second). Their widths range from a few centimetres to a few tens of centimetres (1 to 20 inches). Shadow bands are rather similar to the shadow patterns produced on the bottom of a swimming pool, but much harder to see. They probably arise from a similar effect though, as the last sliver of the Sun is refracted by irregularities in the Earth's atmosphere. They are the inverse effect to the "twinkling" of stars at night. Were stars bright enough visibly to illuminate the surface of the Earth, we would see similar patterns. The movement of the pattern over the pupil of the eye, or the objective of the telescope, results in the brightness variations that we call twinkling or scintillation.

Binoculars and Telescopes

To repeat the warning given in the previous section: observing the Sun during the total phase of the eclipse with the unaided eye, or through binoculars or a telescope is quite safe, and does not need the use of filters or eyepiece projection methods, etc. **However, until the photosphere (the bright solar surface) has been completely obscured, all the safety precautions discussed in Chapter 2 must be applied.**

CAUTION
REQUIRED

Introduction

With binoculars or a hand-held telescope, the instrument can be pointed at the totally eclipsed Sun quickly. However, with a telescope on a mounting it is advisable to have the instrument driven to follow the solar motion whether it is used visually or for imaging (as discussed below), otherwise time will be lost during totality, just setting onto the Sun. For many purposes the normal telescope drive (i.e. the one used for night-time observations) will be adequate. Remember though that the Sun has an intrinsic motion, and therefore for long exposures, long-term tracking, or for critical work, a drive that compensates for the actual motion of the Sun, may be needed. Around the solstices (21 December and 21 June) the Sun's motion in declination is zero, but its right ascension is increasing[72] by about 4 minutes per day. The telescope drive therefore has to be about 0.29% slower than normal. Around the equinoxes (21 March and 21 September), the motion in RA is reduced to 3 minutes 40 seconds per day (drive 0.25% slower than normal), but the motion in declination is about $\pm 24'$ per day, and this will not be corrected by a normal equatorial telescope drive.

Visual Observations

Once the photosphere of the Sun has been obscured by the Moon, and totality has started, observing the eclipsed Sun is identical to normal night-time observing. All the normal night-time observing techniques can therefore be used. Thus binoculars can be used without filters. Telescopes of any size can be used for direct viewing without the use of filters, or the need for eyepiece projection, and so on.

Depending upon the clarity of the sky and the state of the sunspot cycle (Chapter 1), the corona can be from one and a half to three[73] times the size of the Sun, that

---

[72] i.e. it is moving towards the east with respect to the background stars.

[73] This, of course, is just the extent of the corona that can be seen above the background sky illumination. The corona in fact continues right out to where it joins the interstellar gas at a distance of some ten thousand million kilometres (6,000,000,000 miles) or so from the Sun, though decreasing in density and brightness as it does so. Direct observations from spacecraft can show it extending out to 40 or 60 solar radii (10–15°).

is, 0.75° to 1.5° across. A low power, wide angle eyepiece is therefore needed if the whole of the corona is to be seen through a telescope. The field of view of a telescope is given by the eyepiece field of view divided by the magnification:

$$\text{Telescope field of view} = \frac{\text{Eyepiece field of view}}{\text{Magnification}}$$

The field of view of the eyepiece is determined by its optical design, and should normally be specified by the manufacturer. If that information is not available, then the following figures may normally be used: standard eyepiece designs such as the Kellner and orthoscopic have fields of view of 40–50°, the Plössl design has a field of view of 50–60°, the Erfle 60–70° and wide angle designs such as the Nagler up to 85°.

A 200 mm (8 inch) f10 Schmidt–Cassegrain telescope used with a 25 mm (1 inch) eyepiece has a magnification of ×80. Even a wide angle eyepiece will therefore only give about a 1° field of view, and this would often be insufficient to see the whole corona. Such a telescope would need a wide angle eyepiece with a focal length of 50 mm (2 inches: ×40) or a standard type of eyepiece with a focal length of 100 mm (4 inches: ×20) in order to enable the whole corona to be seen with certainty (field of view of 2°).

Features in the corona are often of low contrast, and a low magnification helps to preserve whatever contrast there may be. For coronal work, small telescopes with low magnifications and wide fields of view may well therefore be better than larger, higher power instruments.

The corona usually has a pearly or silvery appearance, but there have been occasional reports of colours within it. Since some of these reports claim to be using averted vision for the observations, and since averted vision puts the coronal image onto parts of the retina with few colour receptors, they perhaps need treating with some scepticism. None the less, trying to estimate any colour cast to the corona either by direct vision, using binoculars, or with a wide angle telescope, should form a part of most eclipse observing programmes (see below).

Observing the prominences can be undertaken with much higher magnifications than those needed for the corona. Prominences range from 10,000 to 100,000 km (6000 to 60,000 miles) in length (Chapter 1), and this corresponds to an angular size of 10″ to 100″, or the equivalent of observing medium to large sized craters on the Moon. The brightness and contrast for prominences is much greater than that of the corona, and they have details on a scale of a fraction of a second of arc. If the atmospheric conditions permit, magnifications of ×250, ×300, or higher may therefore usefully be employed.

The angular rate of motion of the Moon across the sky is about half a second of arc per second of time. Unless they are unusually high above the solar surface, prominences projecting over the leading edge of the Moon will therefore only be visible for about 20 seconds. Similarly prominences projecting past the trailing edge of the Moon will be seen for only about 20 seconds before the end of totality. Successful observing or imaging (see the next subsection) of prominences, wherever they may be around the edge of the Sun, thus requires rapid action on the part of the observer. Prominences to the side of the Moon's track will be visible only if the Moon's angular diameter is no more than 20″ or so larger than that of the Sun. Eclipses with durations of less then a minute therefore often give better views of the prominences than those lasting two or three minutes or longer.

Imaging

Photographic and Digital Cameras

Just as with visually observing the total phase of a solar eclipse, imaging it, or parts of it, proceeds in the same way as the imaging of night-time objects. The same photographic or digital camera, CCD, video camera, etc., may be used directly on the telescope, by eyepiece projection, piggy-backed on the telescope, etc., and again no filter is needed.

The focal length, or effective focal length (Chapter 4) determines the image size, and needs to be optimised for the intended purpose. Thus imaging the whole corona requires a field of view of 1–2°. With 35 mm film, which is actually about 24 mm (1 inch) wide, this means a maximum focal length of about 1 metre (40 inches). If using a CCD with a physical size of 5 mm, the maximum usable focal length would be about 200 mm (8 inches). Conversely using a 35 mm camera with a standard lens of focal length 50 mm

Table 7.2. Fields of view on a 35 mm photographic frame and a 4 mm × 6 mm CCD chip

Focal length	35 mm frame[a]	4 mm × 6 mm CCD
20 mm	70° × 100°	11° × 17°
30 mm	45° × 65°	8° × 11°
40 mm	35° × 50°	5.7° × 8.6°
50 mm	27° × 40°	4.6° × 6.8°
75 mm	18° × 27°	3.0° × 4.6°
100 mm	14° × 20°	2.3° × 3.4°
150 mm	9° × 13°	1.5° × 2.3°
200 mm	7° × 10°	1.1° × 1.7°
400 mm	3.4° × 5.0°	0.57° × 0.86°
600 mm	2.3° × 3.3°	0.38° × 0.57°
800 mm	1.7° × 2.5°	0.29° × 0.43°
1 m	1.4° × 2.0°	0.23° × 0.34°
2 m	0.7° × 1.0°	0.11° × 0.17°
5 m	0.27° × 0.40°	0.046° × 0.068°
10 m	0.14° × 0.20°	0.023° × 0.034°

[a]Very short focal length lenses for 35 mm cameras, sometimes called fisheye lenses, usually have a considerable degree of distortion, and so the actual field of view may differ from the figures given here.

would give an image of the corona just 0.7 mm (0.03 inches) across. A telephoto lens with a focal length of at least 200 mm is thus needed to give a reasonable image of the corona (Table 7.2). If a camera with such a lens is hand-held, then the maximum exposure without camera shake is about 0.01 seconds, and so it will need a focal ratio of f4 or faster (Table 7.3).

Prominences range from a few seconds of arc upwards in size. If you wish to image details of them, then a minimum focal length of 2 metres (80 inches) will be needed whether it is onto photographic emulsion, using a digital camera or a CCD detector. This is because a resolution of 1 second of arc or better will be needed to record the details of prominences, and this corresponds to 0.01 mm (0.0004 inches) for that focal length. Such a physical size is about the same as that of the developed photographic grains or of the pixels of a CCD.

The main differences between night-time and eclipse imaging arise from the brief time available for the observations. The whole total phase of the eclipse is likely to be only a few minutes long, and prominences may only be visible for a few seconds (see the previous subsection). There is therefore no time to allow for focusing the telescope or camera, and little opportunity for optimising the position of a feature in the image frame or for trying a range of exposures. In addition if the telescope, camera, etc., has been used for viewing or imaging the partial phase of the eclipse, then there will be filters to be removed, or other adjustments to be made, before the same system can be used during the total phase of the eclipse.

Cameras and telescopes must thus be accurately prefocused before the eclipse, the required exposures calculated and preset, and as far as possible the instrument directed at the required point in the sky. The latter is not too difficult if images of the whole chromosphere or corona are wanted since the instruments may be set onto the partially eclipsed Sun just before totality (using full-aperture filters, etc., as required). The daily motion of the Sun and Moon take them through their own diameters in two minutes. If you preposition your camera, etc., more than a few seconds before the start of totality you will therefore need to allow for this motion. The instrument should thus be aimed ahead of the actual position of the Sun by half the solar diameter for every minute that you may be ahead of the required time.

It is much more difficult to preset onto prominences, since their positions around the edge of the Sun will only become apparent with totality. If one is available, however, an Hα filter (Chapter 8), can be used prior to the eclipse or during the partial phase to note the positions of suitable prominences in advance of totality.

Prefocusing of telescopes or binoculars may be done on night-time objects if the instrument can then be left set up until the eclipse occurs. If suitable filters (Chapter 2) are available, then with the filters in place, the instrument may be focused on the Sun prior to the start of the eclipse. The filter is then removed for totality. Beware, however, if using metal-on-glass filters, since the focus position may be changed by the filter. If this is the case, then the difference between the focus positions with and without the filter will need to be established by experimentation well before the eclipse occurs. If a camera is being used with a standard or telephoto lens, then the setting will usually be sufficiently accurate to preset the focus.

Instruments may also be focused on distant terrestrial objects. provided that the focal length of the instrument is short. A normal astronomical telescope, with a focal length of a metre (3 feet) or more, however, will not be accurately focused even if the terrestrial object is 1 kilometre (2/3 mile) away. The difference between the focus position for an object a distance D metres away, observed with a telescope of focal length f metres, and for the same telescope used to observe an object in the sky is given by:

$$\text{Inaccuracy in focus position} = \frac{f^2}{D} \text{ metres.}$$

Thus if an object is 500 m (550 yards) away, and the telescope focal length is 2 m (80 inches), the focus setting will be 8 mm (1/3 inch) away from that needed for an object at infinity. If there is no other way of prefocusing, then provided that the distance of the terrestrial object may be measured or estimated, the above equation may be used to calculate the change required in the focal position. The telescope is then focused on the terrestrial object, and the eyepiece moved inwards (i.e. towards the objective of the telescope) by the calculated amount to give the required preset focus. This method, however, is not recommended unless the eyepiece mounting has an accurately calibrated position scale, and even then is second best to the other methods mentioned above.

The exposures needed will vary enormously from one phase of the eclipse to another. Although they can be estimated or calculated in advance, those exposures should always be bracketed by a factor of 2 or 4 on either side. Thus for example if the calculated exposure required is 1/60 second, then you should use exposures of 1/30, 1/60, and 1/125 seconds, or preferably 1/15, 1/30, 1/60, 1/125, and 1/250 seconds. The accuracy of the estimated exposure, and hence the need for extensive bracketing of the exposure, is more critical for slide film than for print film. A reasonable image on a photographic print will still normally be obtained even if the exposure is incorrect by a factor of 4. The autoexposure feature of digital cameras will usually be adequate for most purposes, but may give false settings if the detector is not centred on the part of the image of interest.

The exposures required for the partial phases of the eclipse are the same as for the normal Sun and are given in Table 4.1 for a full aperture metal-on-plastic filter. Do not forget, however, that limb darkening (Chapter 1) causes the edge of the Sun to have only 40% of the intensity of the centre. Although Baily's beads

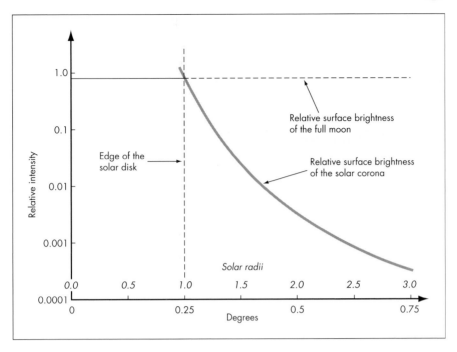

Relative surface brightness
of the full moon

Edge of the
solar disk

Relative surface brightness
of the solar corona

Solar radii

Degrees

Figure 7.8. Typical intensity variations of the solar corona.

and the diamond ring effect still contain portions of the solar photosphere, longer exposures will be needed to show the fainter parts of the phenomena. Indeed imaging the diamond ring should be treated much as imaging prominences (below).

Providing that the observing conditions are similar, and that you use the same equipment, set up in the same manner, the exposure required for the solar corona can be found in advance by experimenting with imaging of the full moon. The corona has a total intensity of about a third that of the full moon at sunspot minimum, and about two-thirds that of the full moon at sunspot maximum. However it is the intensity per unit area, or surface brightness, which determines the required exposure, and this decreases rapidly away from the edge of the Sun (Fig. 7.8). The inner part of the corona therefore needs a similar exposure to that for the full moon, the corona out to 1.5 solar radii needs about 30 times the exposure for the full moon, and out to 2 solar radii, 200–250 times that exposure will be required. A considerable advantage of the variation in the intensity of the corona though is that a very wide range of exposures (Table 7.3) will give usable results. For example at f8, exposures ranging from a hundredth of a second or less to several seconds will all show some parts of the corona. Only if you require the correct exposure for a specific part of the corona will you need to determine a reasonably precise value.

Exposures for the corona may also be estimated from those needed for the Sun. The surface brightness of the Sun is about 400,000 times that of the full moon. A welder's #14 filter reduces intensity by about ×1,000,000 (Table 2.1). The normal Sun, observed through such a filter, may therefore be substituted for the full moon in estimating exposures for the solar corona. The full moon would be about 2.5 times brighter than the Sun through a #14 filter. Be careful to ensure that the filter covers the whole of the objective being used if you estimate exposures in this way, and that the camera viewfinder is also covered by the filter, or blanked off.

Finally the times suggested in Table 7.3 may be used if you cannot estimate the required exposures in any other way, though if you use these figures, they should be well bracketed with a range of exposure times on either side.

CAUTION
REQUIRED

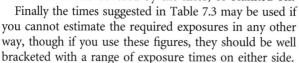

Table 7.3. Exposure guide to the photographic imaging of the solar corona

Focal ratio or effective focal ratio[a]	Exposure times[b] for ISO 100/21° film (seconds)		
	Inner corona (1.1 solar radii)	Corona out to 1.5 solar radii	Corona out to 2 solar radii
2	0.0002–0.001	0.01–0.05	0.05–0.2
4	0.001–0.005	0.03–0.2	0.2–1
6	0.002–0.01	0.06–0.5	0.4–2
8	0.005–0.02	0.2–1	1–5
10	0.008–0.03	0.2–1	1–10
12	0.01–0.05	0.3–2	2–10
14	0.015–0.06	0.5–2	3–20
16	0.02–0.08	0.5–2	4–20
18	0.025–0.1	1–3	5–25
20	0.03–0.1	1–3	5–25
30	0.07–0.3	2–10	10–50
40	0.1–0.5	3–10	20–100
50	0.2–0.8	5–25	40–200
75	0.4–2	10–50	> 50
100	0.8–3	20–100	> 100
150	2–7	50–200	Not possible
200	3–12	> 100	Not possible

[a]See Chapter 1.
[b]For other film speeds, adjust the timings in proportion to the ratio of the first parts of the two ISO numbers; i.e. for ISO 50/18°, the times must be doubled (100/50 = 2), for ISO 200/24° they must be halved (100/200 = 0.5), for ISO 400/27° they must be quartered (100/400 = 0.25) and so on. Exposures longer than about 10 seconds should be increased to allow for reciprocity failure, i.e. the reduction in the speed of response of film at long exposures.

Table 7.4. Exposure guide to the photographic imaging of solar prominences

Focal ratio or effective focal ratio	Exposure times for ISO 100/21° film (seconds)
2	0.0005–0.001
4	0.002–0.005
6	0.005–0.01
8	0.01–0.02
10	0.01–0.03
12	0.02–0.04
14	0.025–0.05
16	0.03–0.06
18	0.04–0.1
20	0.05–0.1
30	0.1–0.2
40	0.2–0.4
50	0.3–0.6
75	1–2
100	1–3
150	3–6
200	5–10

ªFor other film speeds, adjust the timings in proportion to the ratio of the first parts of the two ISO numbers; i.e. for ISO 50/18°, the times must be doubled (100/50 = 2), for ISO 200/24° they must be halved (100/200 = 0.5), for ISO 400/27° they must be quartered (100/400 = 0.25), and so on.

Prominences are somewhat brighter than the inner part of the corona. Their surface brightnesses are less variable than the range from the inner to the outer corona, but none the less do vary from prominence to prominence, and from eclipse to eclipse. If you have determined exposures for the corona from experiments with the full moon, or with the Sun through a #14 filter, then an exposure of 0.1–0.5 that required for the full moon will normally suffice for prominences. But a range of exposures should always be tried. Table 7.4 lists suggested exposures, if you have not been able to determine the values for your instrumentation experimentally.

Most people will want to image both the partial and the total phases of an eclipse. This can cause problems because of the different requirements on the instrumentation and how it is set up between the two phases. If possible it is best to have two completely separate sets of instrumentation, one for imaging the partial phases and another for the total phase. Thus you might have a telescope set up for eyepiece projection for the partial phases, and a camera with a long telephoto lens for the total phase. **NB** Eyepiece projection will not show the total phase of the eclipse at all, so if you are using this method for the partial phases, then some other method is essential for the total eclipse; and remember that there will not be time to rearrange a telescope set up for eyepiece projection to allow direct viewing.

If you intend to use some or all of your instruments for observing both the total and partial phases of the eclipse, then you must have a system whereby you can swap between the two different arrangements required of the equipment as quickly as possible, and there must be a minimum of resetting of the instruments. Ideally there should be no need to refocus, nor to reset the position of the telescope or camera, and the exposures should not change. In practice there is only one way of approaching this ideal. That is by the use of a full aperture metal-on-plastic filter. The filter, provided that it is mounted suitably, can be removed from or added to the telescope or camera very quickly, and if this is done carefully, the position of the instrument will not be disturbed. These filters are so thin that the focus position will be unchanged. Metal-on-glass filters or welder's filters can be used in a similar manner, but you will need to check in advance that the focus position does not change when the filter is added or removed.

One popular method of imaging an eclipse is by multiple exposures onto a single frame. The camera is set up so that it is centred on the position in the sky where the total phase of the eclipse will occur. A series of exposures through a suitable filter will then show the progress of the partial eclipse phases, and the filter is removed to obtain the image of the total eclipse. The Sun and Moon move across the sky at a rate of about $1°$ every 4 minutes, and are each about $0.5°$ across. So exposures at intervals of 4 minutes or more will avoid the successive images overlapping each other. During the eclipse the Sun and Moon will move $30°$ or more across the sky. So a lens with a focal length of 50 mm or less (Table 7.2) has to be used. You will also need a camera that allows multiple exposures to be taken on a single frame. Most modern 35 mm cameras wind on to the next frame automatically and so are not suitable for this purpose. You may, however, be able to pick up an old second hand camera quite cheaply that will allow multiple exposures.

Astronomical CCD Cameras

Imaging the eclipse using a CCD detector is undertaken in the same way as photography, with due allowance being made for the greater sensitivity and smaller size of the CCD. The partial phases of the eclipse may well require exposures shorter than those provided by the cameras software (Chapter 4), and so additional filters or stopping down to quite small apertures may be needed. Imaging the total phase of the eclipse is again similar to imaging the full moon. It can be practised in advance, and the required exposures determined. The small size of most CCD detectors (Table 7.2), means that the pointing accuracy of the telescope or camera is much more critical than it is for 35 mm photography.

Video Cameras

Video cameras do not have the resolution of 35 mm cameras, but do have the advantage of providing a moving image which can capture the "ambience" of the eclipse very well. They can be used as described in Chapter 4, with a full aperture filter for the partial phases, which is removed for totality. Provided that the camera is a modern solid state (CCD) device, the filter can be removed a few seconds before totality to catch the diamond ring (described earlier). Similarly for three or four seconds at the end of the eclipse, the filter need not be replaced. Older video cameras based upon vidicon TV tubes, etc., must not be pointed at the Sun without a suitable filter when there is any portion of the photosphere visible.

Many video cameras will not have lenses that will provide a reasonably sized image of the Sun. A 10:1 zoom lens for example will give an image of the Sun plus corona that is only about a third the width of the TV screen. You may therefore need a converter lens to boost the image size. You will need to practice imaging the Sun with the video camera, filter, and converter in advance of the eclipse because you may find that the automatic focusing system, which often operates using the infrared radiation that is cut out by the solar filter, needs to be disengaged.

Since the whole eclipse lasts for two or more hours, you should also ensure that the batteries of the camera are fully charged, that you have fully charged spare batteries, and that you have several good quality videocassettes to hand. If you record the whole event live, then you will need to change video cassettes several times, it will also take two or more hours to play back the eclipse. If possible therefore it is better to use time lapse to speed up the partial phases of the eclipse. One exposure every second will speed the events up by a factor of 25 or 30. If your camera does not have a time lapse facility, then a similar effect may be accomplished by making very brief normal exposures at intervals of 30 seconds to a minute or so. In either case the camera will need to be on a mounting that is driven to follow the motion of the Sun and Moon across the sky. If you are using a telescope for other types of observation of the eclipse, then the video camera often may be conveniently attached "piggy-back" to the telescope. It is possible to use a standard tripod and to move the camera to centre on the Sun between each exposure, but unless this is done very accurately the images will jump around unpleasantly on the screen.

Video cameras may also be used in a normal mode to record the scene at the eclipse site. This can form a valuable record of what you actually did, because you may forget to record everything in the hectic minute or two of the total phase. But it may also show other phenomena such as the approach and recession of the Moon's shadow before and after totality.

Shadow Band Imaging

Shadow bands (Described above under "Naked Eye Observing") are difficult to image because of their low contrast. They are most easily seen on white or very pale surfaces. If there is not a conveniently painted wall available, then large sheets of white paper or cardboard will need to be set up. Still or video photography is then undertaken as normal looking at the white surface, but exposures will need to be precisely correct if any bands are to be detected.

Observing Programme

It is essential to plan and to practise your observing programme well in advance of the eclipse. The details of the observing programme will depend upon your interests, the equipment available, whether it is your first or twenty-first eclipse, the nature of the observing site and the facilities available there, and on whether you are working on your own or as a part of a team, etc. The programme is necessary to ensure that you make all the observations and obtain all the images that you want. It is also needed so that you can make sure that one activity or set of equipment does not interfere with another. Finally it is needed so that you can make sure that you have all the required equipment, such as cameras, lenses, binoculars, telescopes, filters, film, batteries, video cassettes, etc., available and in working order, and that you have determined all the exposures and focus settings that will be needed. **NB** One item to be taken along to a total eclipse session which would probably not normally be considered is a torch. It can be sufficiently dark during totality that you will not have sufficient ambient light to read telescope and camera settings, etc.

Because most people, even those who go on several eclipse expeditions, succeed in seeing total eclipses on only a few occasions, there is a great tendency to try to cram as much activity into the brief interval of totality as possible. Thus several cameras may be set up on telescopes or with various lenses or with special filters, CCDs, videos cameras, or even spectroscopes can be used. You may, however, then become so occupied with all this equipment, that you never really **see** the eclipse at all. It is recommended therefore that you do not attempt to do too much, but leave yourself time to enjoy and watch the progress of the eclipse in peace. Even if you do intend to zoom around amongst many instruments taking all types of images, you should plan a programme that occupies only half the time of the total phase of the eclipse. This is to allow the opportunity to correct those things that go wrong, or that do not go quite as you had planned. You can always have a few additional items in a backup programme to allow for the unlikely possibility of everything working perfectly, and so finding your main programme completed half-way through the eclipse.

Whatever your observing programme for the eclipse may be, it is essential to plan it thoroughly, and to practise it in advance. Otherwise you are likely to waste time, to find that one activity interferes with or overlaps another, to make mistakes, and worst of all perhaps end up with no results at all.

Specialist Instrumentation

Just as with direct solar observing (Chapter 8), there are a number of specialist instruments designed for use during solar eclipses. Most of these, however, are available only at large observatories and to professional astronomers. There are one or two though which can be considered for amateur use

Radial Density Filters

The surface brightness of the solar corona decreases very rapidly with increasing distance from the Sun (Fig. 7.8). Even CCDs which have large dynamic ranges are unable to cover the whole of the corona. An image therefore which is correctly exposed for the inner corona will show the outer corona only very weakly, or not at all. While an image showing the outer corona will have the inner part over-exposed. This problem may be overcome through the use of a radial density filter. This is a filter whose optical density decreases away from its centre. When superimposed on the image of the corona, it reduces the intensity of the inner parts of the corona in comparison with the outer parts and allows a photograph or CCD image to be obtained which shows details of all parts of the corona.

An ideal radial density filter would have its optical density smoothly decreasing outwards from its centre. While such filters can be made to order by specialist optics companies, they would be impossibly expensive for most people. A reasonable substitute may be made by photographing the image shown in Fig. 7.9 onto black and white film and with the camera slightly out of focus so that the dots are smeared out. The negative of this image will then give the required filter. You will need to calculate or measure the size of the solar image (**caution** – use all the precautions discussed in Chapter 2 whenever you observe the Sun), and use a camera lens that will match the size of the central part of Fig. 7.9 to that image. In use the filter is placed just in front of the photographic emulsion or CCD chip, and aligned accurately with the centre of the Sun. An exposure suited to the outer parts of the corona will then show details across most of the corona.

Figure 7.9.
Photographing this image onto black and white film, and with the camera slightly out of focus, will produce a radial density filter for coronal imaging. The camera will need to use a lens whose focal length matches the size of the central part of this image to the size of the image of the Sun in the telescope.

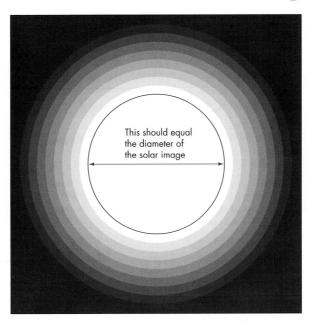

This should equal the diameter of the solar image

Another substitute which will still give reasonable results if used carefully is a step filter which is positioned a little way from the focal plane of the telescope. The step filter can be easily made by a DIY enthusiast. You will need a neutral density acetate-based filter with an optical density of 0.2D. This is obtainable to order from many photographic shops. A single layer of this filter will reduce the intensity of the light to 63%. Two such filters on top of each other will reduce the intensity to 40% (63% of 63%). Three layers will reduce the intensity to 25% and so on. If the diameter of the image of the solar disk at the focus of your telescope is then x mm, you will need to cut circles of the following diameters from the filter; 1.1x, 1.2x, 1.3x, 1.4x, 1.5x, 1.6x, 1.7x, 1.8x, 1.9x and 2.0x. Superimposing and aligning all the circles cut from the filter will then give a step radial filter which varies in optical density from 2.0D (a reduction in intensity by a factor of 100) at the centre to zero outside the last filter.

An alternative to a radial density filter is to use image processing (Chapter 4) to combine images with differing exposures. These can either be CCD or digital camera images, or photographic images which have been scanned onto a computer disk. Saturated or under-exposed parts of each image can be deleted using the image processing software, and then the correctly exposed parts of each image co-added. Smoothing, unsharp masking, grey-scaling, contrast and brightness adjustment, etc., will need to be used to get a final composite image showing details of all parts of the corona.

Slitless Spectroscopes

The solar chromosphere is the lowest layer in the solar atmosphere (Chapter 1) and is visible as a thin red ring just above the photosphere for a few seconds at the start and end of totality. The chromosphere and prominences produce emission line spectra (Chapter 8). A simple low dispersion spectroscope used without an entrance slit will produce a series of images of the chromosphere and of any prominences in the light of each of their emission lines. This is known as the flash spectrum (Fig. 7.10) because it flashes into view for a few brief moments at the start and end of the total phase of the eclipse.

The flash spectrum can also be observed through the use of an objective grating or an objective prism. These are large coarse transmission diffraction gratings or large low angle prisms placed in front of the objective of the telescope. The image of anything seen through the telescope is then replaced by its spectrum. An objective prism is unlikely to be available to most people, but an objective grating can be produced if you have access to

Figure 7.10. The flash spectrum of the solar chromosphere (photograph reproduced by permission of the Royal Astronomical Society).

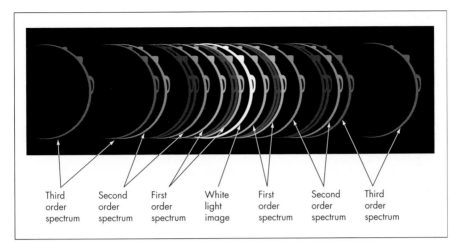

| Third order spectrum | Second order spectrum | First order spectrum | White light image | First order spectrum | Second order spectrum | Third order spectrum |

Figure 7.11.
Schematic flash spectra produced by an objective grating with lines at 0.1 mm (0.004 inch) intervals.

a computer drawing package and a good quality printer. The printer will need to be able to print onto clear acetate sheets, or you will need to get the paper output from the printer photocopied onto an acetate sheet. The objective grating is then simply a series of parallel black lines each 0.05 mm[74] (0.002 inches) wide separated by 0.05 mm (0.002 inches) clear gaps, and covering an area large enough to go over the telescope. The acetate sheet should be mounted to keep it as flat as possible without stressing it. A simple transmission grating like this will produce a direct (white light) image in the centre of the field of view, flanked on either side by several spectra. Since the light is split up into many images, only the brightest emission lines are likely to be visible (Fig 7.11).

Transits

CAUTION
REQUIRED

Transits are similar to solar eclipses in the geometry of the situation, but differ in that the intervening body between the Earth and the Sun has a small angular size. Although in theory there could be transits of asteroids and comets, it is only Mercury and Venus that are of real concern. When they occur, transits are normally visible from the whole Sun-lit half of the Earth. Only a tiny fraction of the Sun's disk is obscured by the planet,

---

[74] These measurements are not critical – anything near these figures will work.

Table 7.5. Transits of Mercury and Venus

Mercury	1999 15 November
	2003 7 May
	2006 8 November
	2016 8 May
Venus	2004 7 June
	2012 5 December

so that observing transits is equivalent to observing the Sun directly, and all the precautions discussed in Chapter 2 must be applied.

As with solar eclipses which occur when the Sun is near the nodes of the Moon's orbit, transits can only occur when the Earth–Sun line is close to the lines of nodes for the orbits of Mercury and Venus. As a result transits are much rarer than eclipses. Transits of Mercury occur on or about 7 May or 9 November, and at intervals of a few years or a few tens of years. Transits of Venus occur around 7 June and 8 December in pairs separated by eight years. Each pair of transits is then separated from the next by a 130-year interval. Table 7.5 lists the next few transits for both planets,

Transits of Mercury and Venus were of interest historically as a means of measuring the distance from the Earth to the Sun,[75] and for studying the atmosphere of Venus. Nowadays they are of interest primarily for their own sake.

[75] Captain James Cook's first round-the-world voyage had as one of its aims the observation of the 1768 transit of Venus.

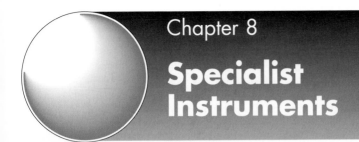

Chapter 8

Specialist Instruments

The Sun is so bright and so angularly large that there are many specialist instruments that have been developed specifically for observing it in different ways. Some of these instruments are potentially affordable to an amateur astronomer able to purchase a 0.2 m (8 inch) Schmidt–Cassegrain telescope, others may be potentially constructable by a keen DIY enthusiast. Most such instruments, however, must be exclusive to a large, well-funded research team (Chapter 10).

Narrow Band Filters

Hα Filters

The filters considered in Chapter 2 reduce the solar radiation to safe levels by having very low transmissions across the whole of the optical part of the spectrum. A different approach is to use a filter that has a comparatively high transmission ($\simeq$10%), but only over a very small range of wavelengths (colours). The range of wavelengths, or band width, of such a filter must be about 1 nm[76] or less for safety. Many actual filters have band widths down to 0.1 nm or even 0.05 nm.

From purely the safety viewpoint, the wavelength on which the narrow band filter is centred could be

---

[76] The visible part of the spectrum extends from about 380 nm (violet) to 700 nm (red), so 1 nm represents about 0.3% of the visible spectrum.

anywhere across the visible spectrum. In practice such filters are always centred on one of the Sun's strong spectrum lines (see the section below on spectroscopy), and in most cases on the red line arising from hydrogen. This line is known as the Balmer Hα line and is the first of a series of lines progressing towards the violet. Its wavelength is 656 nm. The next lines in the series are Hβ at 486 nm (blue-green) and Hγ at 434 nm (blue). The reason for centring the filter on a strong spectrum line is that solar features such as prominences, filaments, flares, etc. (Chapter 1) may be most easily seen. Such features do not emit their radiation across the entire spectrum as does the Sun as a whole, but only at the wavelengths of hydrogen's (and a few other elements') spectrum lines (Fig. 7.10). The prominence, etc., is therefore reduced in intensity far less than the rest of the Sun by a narrow band filter and so becomes more readily observed because of its increased contrast against the disk of the Sun.

Hα filters, and other similar filters operating at other wavelengths, are often called interference filters, because they operate through light waves interfering with each other. The term interference is here used in a technical sense to mean different light waves adding together to reinforce their intensity, or cancelling each other out to reduce or eliminate their intensity. The filter is designed so that light waves of the required wavelengths add together and pass through the filter, while all other wavelengths are rejected.[77] The filters are based upon a partially reflecting mirror. This then has a thin layer of a substance such as magnesium fluoride deposited onto it and a second partially reflecting mirror placed on top of that (Fig. 8.1). The light paths through the filter are shown in Fig. 8.2. The separation of the partially reflecting mirrors is chosen so that the light beams emerging on the right in the figure add together at the wavelength of the Hα line, while those reflected to the left cancel each other out.

The central wavelength of the filter can be changed by changing the thickness of the centre layer. The layer would be around 100–200 nm thick for an Hα filter. The band width can be changed by changing the percentage of light reflected by the two partially reflecting mirrors. The whole filter has to be manufactured and main-

---

[77] The filters are also known as Fabry–Perot etalons. The complete theory of their operation may, for example, be found in *AT*.

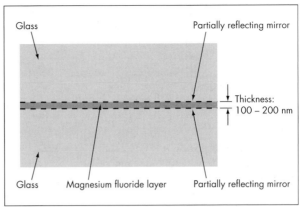

Figure 8.1. The construction of an interference filter.

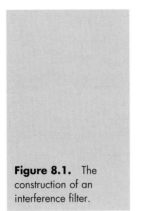

tained to an accuracy of a few tens of nanometres. They are usually mounted in a small heated box (Fig. 8.3) and in use maintained at a precise temperature to ensure that this level of accuracy is achieved.

The effective thickness of the central layer is changed if light passes through the filter at an angle. If the filter is just 5° from being square-on to the light beam, then the effective thickness of the layer will be increased by 0.4%. This is sufficient to change the wavelength transmitted by the filter from 656 nm to nearly 659 nm, and so for it to miss the Hα line completely.

Figure 8.2. Light paths within an interference filter.

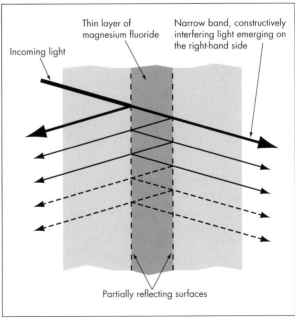

Figure 8.3.
a An Hα filter mounted onto a 180 mm (7 inch) Maksutov telescope; **b** note that the telescope is stopped down to 75 mm (3 inches) by an additional red filter before the aperture of the telescope.

The filters have therefore to be mounted carefully onto the telescope. In the case of the filter shown in Fig. 8.3, the use of the entrance filter means that the light beam entering the telescope is off-axis. The Hα filter thus has to be mounted onto a low angle wedge so that it is orthogonal to the actual light beam emerging from the telescope.

For some filters, the change in wavelength with angle to the light beam is used to provide fine tuning for the central wavelength of the filter. The filter is designed to pass light of a slightly shorter wavelength than that required, and then its angle to the light beam may be adjusted to allow the emerging light to centre on the Hα line, or to scan through it.

The dependence of the filter wavelength upon the angle that the light passes through the filter means that ideally the light beam passing through the filter should be a parallel one. In practice, beams from telescopes are converging since the telescope is focusing the parallel light from the Sun. A 5° convergence angle corresponds to an f-ratio of about f6, and this would lead to the filter having a pass band 2.6 nm wide. Such a wide pass band would then lead to very little detail being seen in the resulting solar image. Telescopes therefore have to be stopped down to focal ratios of f20 or longer if the filter is to operate within its designed band width. Even at f20, the band width of the filter is increased by 0.2 nm. The reduction of the effective focal ratio is thus the primary purpose of the entrance filter shown in use in Fig. 8.3.

Hα filters are available commercially from a number of suppliers who advertise in the popular astronomy magazines (Appendix 2). They are, however, expensive. The typical cost would be £1000–2000 ($2000–3000) at the time of writing, for filters with a bandwidth of 0.1 nm, and considerably more for narrower band-widths. As may be seen from the description of their mode of operation, they are not suitable for DIY construction. A device which is just within the capabilities of a very well-equipped and skilled DIY enthusiast, and which also produces a narrow band image of the Sun, is the spectrohelioscope,[78] discussed later. The actual cost of building such an instrument, however, including labour, is likely to be many times that of a commercially produced Hα filter. Against the cost of the Hα filter (or spectrohelioscope), however, must be set the opportunity to view features of the Sun not observable by any other means (Chapter 1), and so the keen solar observer may well consider the expenditure worthwhile.

---

[78] Prominences, but not filaments or flares, may be observed using a much simpler instrument – the prominence spectro-scope – described below.

Lyot Monochromators

The Hα filters described above rely upon the interference effects of light for their operation. Another type of narrow band filter is based upon the combined effects of birefringence and interference and is known as a Lyot monochromator. Birefringence is the phenomenon whereby in some crystals, notably quartz, two light beams polarised at right angles to each other travel through the crystal at different velocities. With a crystal of the right thickness, therefore, one beam can be arranged to have gained half a wavelength on the other (180° out of phase), and therefore the two beams interfere destructively. By combining a number of quartz crystals of varying thicknesses, a filter can be produced with any desired bandwidth at any desired operating wavelength. Like the Hα filters previously described, however, Lyot monochromators are usually designed to centre on one of the strong absorption lines in the solar spectrum, mostly Hα at 653 nm or the violet lines due to ionised calcium at 397 nm (Ca H) and 393 nm (Ca K). Lyot monochromators are only produced to order by specialist optics suppliers; they are many times the price of the interference type filters, and unsuited to amateur use.

Spectroscopy

Introduction

A great deal, possibly the majority, of our knowledge of objects in the sky comes from the study of their spectra. The Sun is no different from other objects in this respect, except that as with other types of observation, the enormous amounts of energy available make it possible to use specialised solar instruments.

Spectroscopy is the technique of splitting the light from the Sun, or other object, up into its component colours and spreading these out in order of increasing or decreasing wavelength. It can be carried out in any part of the total spectrum, from gamma rays through X-rays, the ultraviolet, optical and infrared to the microwave and radio regions. However, it is only in the visual part of the spectrum that anyone other than a specialist research team is likely to be able to build and operate a spectroscope.

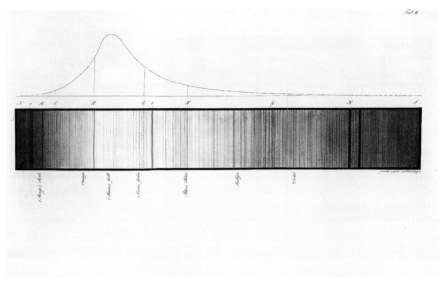

Figure 8.4. The visual solar spectrum. Drawing made in 1823 by Joseph von Fraunhofer (1787–1826). Reproduced by permission of the Royal Astronomical Society.

In the visual region, the solar spectrum is familiar to everyone from rainbows. This, however, is not a pure spectrum, and the differing colours overlap and blur together. With a properly designed spectroscope, the dark, and occasionally bright, lines present in the spectrum become visible (Fig. 8.4).

The spectrum lines arise from the interaction of the light emitted by the inner parts of the Sun's atmosphere with elements in its outer atmosphere. The atoms of an element absorb light at specific wavelengths in a pattern which is unique to that element[79] (Fig. 8.5). By recognising those patterns it is possible to identify which elements are present in the object being observed. Thus in Fig. 8.4 the lines labelled, C, F and G are the Hα, Hβ and Hγ lines due to hydrogen, and demonstrate its presence in the Sun. Similarly the line in the yellow marked D (which is actually a double line) comes from sodium. When sodium emits light, as in the ubiquitous sodium street lamps, it does so at the same wavelength as these absorption lines and produces the familiar yellow glow.

In this way we find that the Sun is made up from the same elements that are found on Earth, but in rather different proportions (Table 8.1).

Furthermore, if an atom loses an electron and becomes ionised, then the pattern of lines in its

[79] For further details of the atomic basis of spectroscopy see, for example, *OAS*.

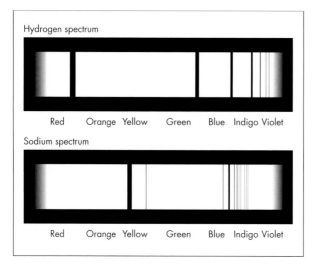

Hydrogen spectrum

Red   Orange Yellow   Green   Blue Indigo Violet

Sodium spectrum

Red   Orange Yellow   Green   Blue Indigo Violet

Figure 8.5.
Schematic spectra of
hydrogen (top) and
sodium (bottom)
showing the main
absorption lines of each
element, and their
differing patterns.

spectrum changes to a new and different unique form. Loss of a second or third electron (double and triple ionisation, etc.) produces yet further patterns. Knowing the temperatures at which atoms will ionise therefore provides a measure of the temperature of the object being observed. Thus lines due to singly ionised iron can be found in the spectrum of the normal solar surface which has a temperature of about 5800 K, but lines from iron atoms which have lost 12 or more electrons are found in the spectrum of the solar corona (Chapter 7), and indicate its temperature to be around 1,000,000 K.

Molecules also produce spectrum lines, although these are more usually to be found at microwave and

Table 8.1 Abundance of the main elements in the Sun by mass

Element	Abundance (%)
Hydrogen	73.49
Helium	24.90
Oxygen	0.77
Carbon	0.29
Iron	0.16
Neon	0.12
Nitrogen	0.09
Silicon	0.07
Magnesium	0.04
Sulphur	0.04
Argon	0.02
Nickel	0.01
All other elements	<0.01

radio wavelengths. For the Sun, spectrum lines from titanium oxide (Ti O) may be seen in the optical spectra of sunspots, showing their temperatures to be down to 4000 K or so.

The absorption lines from an element are produced at precise wavelengths. Thus the first few lines in the visible spectrum of hydrogen are at:

Hα 656.2808 nm
Hβ 486.1342 nm
Hγ 434.0475 nm
Hδ 410.1748 nm

These values are as they would be observed in the laboratory. But astronomical objects are often moving along the line of sight with respect to the Earth, and this changes the wavelengths through the Doppler shift (Chapter 1). Thus careful observation of (say) the Hα line shows that on one edge of the Sun it has a wavelength of 656.2764 nm whereas at the opposite edge the wavelength is 656.2852 nm. This difference arises through the rotation of the Sun which leads to the one edge approaching us and the other moving away. Comparing these observed wavelengths with the laboratory wavelength of 656.2808 nm leads to a rotational velocity for the Sun at its equator of 2 km/s, and so to a rotation period of about 25.5 days.[80]

There is insufficient space in this book to go further into the details of how and why spectroscopy forms such an important tool for the astronomer. Readers wishing to pursue an interest in spectroscopy further are referred for example to the author's *OAS*.

A Spectroscope

A simple spectroscope can be made by any reasonably skilful DIY enthusiast. The main item required is the device which splits the light up into the various colours. For a small spectroscope, this will be either a prism or a diffraction grating, and these can be purchased at fairly low cost from suppliers of equipment to schools,

---

[80] The Sun does not rotate like a solid body, so this is the rotation period at the equator. The rotation period increases away from the solar equator to 30 days or more near the poles (Chapter 5).

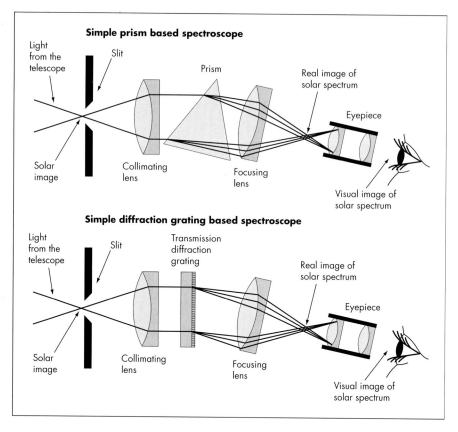

Figure 8.6. A simple prism spectroscope (top) and a simple diffraction grating spectroscope (bottom).

sometimes from army-surplus stores, or from telescope manufacturers. The other items needed will be a slit to form the entrance aperture of the spectroscope, which will typically be 0.05–0.25 mm (0.002–0.01 inches) wide, a couple of lenses and an eyepiece. The spectroscope layout is then shown in Fig. 8.6.[81]

The entrance slit to the spectroscope takes the place of the eyepiece in the telescope, and the solar image is focused onto it. The slit is positioned at the focus of the collimating lens, so that a beam of parallel light emerges from the latter. That beam is then directed onto the prism or transmission diffraction grating which splits it into the various colours. The imaging lens produces a real image of the spectrum, which may be photographed or imaged with a CCD if wished. Alternatively an eyepiece placed after the real image will provide a visual image to the eye.

---

[81] For further details see *OAS* and *AT*.

CAUTION
REQUIRED

Spectroscopy of stars and galaxies requires the largest of telescopes, but for the Sun quite small telescopes are adequate. Since the light from the Sun is spread over many wavelengths in a spectrum, there is less intrinsic danger in observing solar spectra than the direct solar image. None the less, precautions need to be taken, and the white light image produced before entry into the spectroscope is still as damaging to eyes or equipment as that produced for direct viewing. The telescope should be stopped down to 20 or 30 mm (about 1 inch). The criteria for safe viewing of the solar spectrum directly are then complex and depend upon the spectroscope design as well as that of the telescope. However, the same intensity limits apply as for direct viewing; a reduction to less than 0.0032% of the direct intensity in the visible region of the spectrum (Chapter 2). For a spectroscope based upon a small 60° prism or a blazed transmission grating[82] used in the first order with 250–500 lines per mm, the requirements for safe viewing are:

$$\frac{F_{\text{Collimator}} \times F_{\text{Eyepiece}}^2 \times \text{Exit beam width}^2 \times \text{Slit width}}{F_{\text{imaging lens}}^4 \times \text{Telescope EFR}^2} \leq 10^{-14}$$

where
$F_{\text{Collimator}}$ is the focal length of the spectroscope collimating lens;
F_{Eyepiece} is the focal length of the spectroscope eyepiece;
$F_{\text{Imaging lens}}$ is the focal length of the spectroscope imaging lens;
Exit beam width is the width of the light beam coming out of the prism;
Slit width is the width of the entrance slit of the spectroscope;
Telescope EFR is the effective focal ratio of the telescope (i.e. based upon its stopped-down diameter).
NB All linear measurements must be in the same units.

Thus for a 0.2 m (8 inch) f10 Schmidt–Cassegrain telescope which has been stopped down to 25 mm (1 inch) used with a spectroscope with a slit width of 0.2 mm (0.008 inches) a collimating lens of focal length, 50 mm (2 inches), a 20 mm (0.8 inch) prism, an imaging lens with a focal length of 200 mm (8 inches) and an eyepiece of focal length 25 mm (1 inch), we have

$$F_{\text{Collimator}} = 0.05 \text{ m}$$
$$F_{\text{Eyepiece}} = 0.025 \text{ m}$$

---

[82] See *AT*.

$F_{\text{Imaging lens}} = 0.2$ m
Exit beam width $= 0.02$ m
Slit width $= 0.0002$ m
Telescope EFR $=$ f80 (focal length $= 2$ m,
stopped-down diameter $= 0.025$ m)

and so

$$\frac{F_{\text{Collimator}} \times F^2_{\text{Eyepiece}} \times \text{Exit beam width}^2 \times \text{Slit width}}{F^4_{\text{Imaging lens}} \times \text{Telescope EFR}^2}$$

$$= \frac{0.05 \times 0.025^2 \times 0.02^2 \times 0.002}{0.2^4 \times 80^2} = 2 \times 10^{-13}$$

The image in this telescope–spectroscope combination would be too bright by a factor of five for direct viewing. A safe image intensity could be achieved by stopping the telescope down to 11 mm (EFR = f180), by changing to a 10 mm eyepiece, by decreasing the slit width to 0.04 mm, or by stopping down the prism (exit beam width) to 9 mm, etc. It could also be achieved by changing to an imaging lens with a focal length of 300 mm, but that is less practicable, since it would probably mean re-building the spectroscope. In practice this calculation will err on the side of safety by about a factor of two because of additional light losses by reflection and scattering at the numerous glass surfaces involved in the spectroscope.

Prominence Spectroscope

A simple spectroscope such as those shown in Fig. 8.6 can be made by a competent DIY enthusiast and will suffice to show the main lines in the solar spectrum. A much more sophisticated instrument, however, would be needed to enable velocities to be determined, and a good deal of ancillary equipment, computing power and reference books as well as specialist knowledge would be needed to attempt to determine the solar abundances of the elements. Such types of work probably therefore remain the province of the professional astronomer. A simple spectroscope, however, can be used in a manner which takes advantage of the properties of emission line spectra[83] to enable promi-

---

[83] As previously noted, all elements absorb light at precise wavelengths which are unique to each element, molecule, ion, etc. The absorption occurs when the element, etc., in the form

nences (Chapters 1 and 7) to be observed. This is through its use as a prominence spectroscope.

Prominences are condensations in the lower part of the solar corona (Chapter 1). They are often linear in structure and can be up to 100,000 km long. Their shapes frequently suggest that they are influenced by local magnetic fields, and loops are quite a common form for the smaller prominences. The gas within prominences has a temperature around 10,000 K and so is considerably hotter than the visible solar surface. The density of the prominence material is, however, much less than that of the surface, and so when seen across the disk of the Sun, prominences appear as dark silhouettes. They are then called filaments, and can be observed using Hα filters, monochromators or a spectrohelioscope (all of which are discussed in other parts of this chapter).

Prominences are some 10,000 km above the visible solar surface, and so project out from the edge of the solar disk, often looking rather like huge flames leaping out into space. In theory, therefore, it should be possible to see prominences by any of the methods of observation discussed in earlier sections. In practice the prominences are then completely lost in the glare from the very much brighter white light image of the Sun. But we can observe prominences by taking advantage of the different types of spectra produced by the Sun and the prominence. The Sun, as we have seen, has a spectrum with a bright continuous background and dark absorption lines. The prominence is a mass of hot gas seen against a dark background, and therefore produces an emission line spectrum. The light from the prominence is thus concentrated into a few emission lines and not spread across all wavelengths (Fig. 7.10). If we select one of those emission lines (usually Hα at 656 nm), and devise a means of observing the Sun at just that wavelength then prominences will be easily

---

(footnote 83 continued)
of a gas, is silhouetted against a bright background. Thus the outer layers of the Sun, projected against the bright background of the deeper layers, produce the Sun's absorption spectrum. When a hot gas is projected against a dark background, as for example with prominences extending beyond the solar limb, bright lines are seen in the spectrum against a dark background. That is an emission line spectrum, and the emission lines occur at the same wavelengths that the elements, etc., would absorb radiation in an absorption line spectrum.

seen. The prominence becomes visible because the contrast between the disk of the Sun and the prominence is greatly reduced by observing at the chosen wavelength. There are several reasons for this change in contrast; firstly most of the white light from the solar disk is eliminated, secondly the solar disk is dimmed by observing at a wavelength of one of its absorption lines, and finally the prominence concentrates its energy into the emission line. The Hα filter and spectrohelioscope also enable prominences to be seen by this principle, but the prominence spectroscope provides a simpler and cheaper means to the same end.

The prominence spectroscope is a small spectroscope adapted to fit in place of the eyepiece of a telescope which has been stopped down to 50–75 mm (2–3 inches). **Caution** – although the final image will be safe to see, remember and apply all the precautions required for setting a telescope on the Sun discussed in earlier sections. The prominence spectroscope has a wider than normal entrance slit, and a second slit superimposed on the spectrum over the selected spectrum line. An eyepiece then provides a view through this second slit (Fig. 8.7). The entrance slit is aligned so that it covers the edge of the solar disk and thus the spectroscope produces an image of it and of any prominences in the region in the light of the selected emission/absorption line.

CAUTION
REQUIRED

Since prominences come and go through their own changes and through the rotation of the Sun, it will usually be necessary to sweep around the edge of the solar disk with the spectroscope entrance slit in order to find them. This can be done using the telescope slow motions. However, a convenient way of operating the prominence spectroscope is to mount it so that it is displaced to the side of the optical axis of the telescope to the point where the entrance slit aligns with the edge of the solar image when the telescope is centred on the Sun. By rotating the spectroscope mounting, the entrance slit will then scan around the edge of the Sun quickly and simply.

Spectrohelioscope

The spectrohelioscope is closely related to the prominence spectroscope, but is far more stringent in the

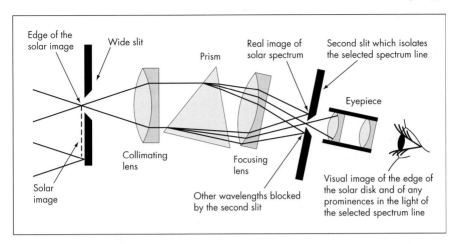

Labels in figure:
Edge of the solar image
Wide slit
Prism
Real image of solar spectrum
Second slit which isolates the selected spectrum line
Eyepiece
Collimating lens
Focusing lens
Solar image
Other wavelengths blocked by the second slit
Visual image of the edge of the solar disk and of any prominences in the light of the selected spectrum line

Figure 8.7. The prominence spectroscope.

requirements for its construction. None the less, the device is constructable by a good DIY enthusiast.

The optical principles of the spectrohelioscope are identical to those of the prominence spectroscope (Fig. 8.7) except that a normal narrow entrance slit is used. The instrument provides an image of the whole Sun at the selected wavelength, however, not just of the edge. This is accomplished by moving the entrance slit of the spectroscope so that it scans across the solar image produced by the telescope. The movement of the entrance slit causes the spectrum to move, and so the second slit has to follow that movement in order to remain aligned with the selected spectrum line. If the oscillations of the slits are at a rate of 10–15 Hz or more, then the eye will see a continuous image of the whole Sun in the light of the selected spectrum line. Slower scanning rates can be used and images produced, if the eye is replaced by a photographic, digital or CCD camera.

The mechanical constraints on the construction of the moving slits of the spectrohelioscope are very tight if they are to remain in correct alignment. This is the main problem in constructing the device. Once working, however, it provides a monochromatic image of the Sun like that produced by the Hα filter and the Lyot monochromator. Unlike those devices, however, it is easily reset to spectrum lines other than Hα by moving the second slit along the spectrum. Its pass band can also be changed simply by changing the widths of the slits. It is thus a much more flexible device than the filters, and can easily be optimised for different purposes and types of observation.

Coronagraph

A solar eclipse (Chapter 7) occurs when the Moon passes in front of the Sun. The corona and other features become visible during totality because the very much brighter photosphere whose scattered light normally swamps them is obscured. In theory any means of blocking out the solar photosphere should have the same effect. The production of an artificial eclipse would have many advantages – including the ability to observe the corona at will and from a convenient site, not just from when and where the capricious "Gods" of orbital mechanics dictate that an eclipse will take place.

In practice producing a successful artificial eclipse is very difficult. The main problem is that the solar photosphere is a million times brighter than the corona. So if even 0.001% of the photospheric light fails to be removed by the artificial eclipse, the corona will be overwhelmed by that remnant of scattered photospheric light. None the less, devices have been constructed that allow at least the inner part of the corona to be seen via an artificial eclipse. These instruments are called coronagraphs.

The basic principle of a coronagraph is straightforward; a telescope produces an image of the Sun, the photospheric part of that image is then blocked out by a stop, and the light from the corona is then re-imaged. In order to work, coronagraphs have to be designed to minimise the scattering of photospheric light. The main precautions required are:

- placing the instrument at a high altitude site to minimise scattering in the Earth's atmosphere;
- the use of a simple converging lens (i.e. not an achromat) as free as possible from bubbles, straie and other imperfections in the glass, for the telescope objective, so that reflection or scattering from multiple glass surfaces is minimised. A refractor is preferred because most coatings used to produce mirrors scatter a small proportion of the light as well as reflecting the major portion. Reflecting telescopes are thus less well adapted to being used as coronagraphs;
- the use of extra stops and apertures to eliminate diffracted or multiply reflected light from the photosphere;

Figure 8.8. The solar corona imaged with a coronagraph (photograph reproduced by permission of the Royal Astronomical Society).

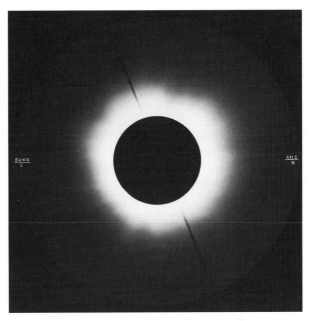

- keeping all optical surfaces and light paths scrupulously clean and dust-free.

With these precautions and under good observing conditions it is then possible to observe the corona. Figure 8.8 shows such an image of the corona taken with a coronagraph.

Solar Telescopes

Telescopes specifically designed for observing the Sun differ in several ways from conventional night-time instruments (see also Chapter 2). The single most important factor in getting good solar images is the site. This must minimise atmospheric turbulence. The differential heating of the dome, telescope and its surroundings by the Sun leads to strong convection currents which degrade the image quality. Attempts to minimise these convection currents include siting the telescope in the middle of a lake, planting the surrounding area with shrubs and small trees, and siting the telescope objective high above ground level. The last solution results in tower telescopes wherein the telescope objective is fixed at the top of a tower some ten or twenty metres (30–60 feet) high. The light from the Sun is then fed into the (fixed) telescope using a coelostat.[84]

Parts of the telescope itself can also be heated by the Sun, and can lead to convection within the instrument. To overcome this turbulence, the telescope may be sealed and filled with helium. The high conductivity of helium then minimises the convection currents. Ultimately, the interior of the telescope may be evacuated to produce a vacuum telescope with no internal convection currents at all.

Any design of telescope can be used for solar observing, but the specialist solar telescopes are often refractors. This is for two reasons: firstly, it is easier to produce a sealed system for a helium-filled or vacuum telescope and, secondly, the refractor does not have a secondary mirror to diffract and scatter the solar radiation in the manner of reflecting telescopes. The focal ratios of solar telescopes are often very high, so that a large image of the Sun is produced without the need for secondary optics, which would add to the light scattering.

Adaptive optics is currently being tried out as a means of further improving the images produced by solar telescopes. This is a method of reducing the distorting effect of atmospheric turbulence. It relies upon reflecting the light beam from a small flexible mirror. The mirror is mounted on a number of supports, whose positions can be quickly altered under computer control. The atmospheric distortions are measured, and then the flexible mirror is distorted in an equal and opposite manner.[85] The resulting images can be a great improvement over those from conventional telescopes, but the method requires a small dark sunspot to be present in order to measure the atmospheric distortions, and the corrected image may be only a few seconds of arc across. So it is not a universally applicable technique.

---

[84] A coelostat is a device for feeding the light from the changing position of the Sun (or other object) in the sky into a fixed direction. There are several designs, but most use two flat mirrors, both of which are driven to produce the fixed light beam. They are essential for large, cumbersome instruments which cannot themselves be moved to track the Sun.

[85] For further details see, for example, *AT*.

Chapter 9

Radio Telescopes

If we had eyes sensitive to radio waves at a frequency of 100 MHz,[86] then the sky would usually contain three main objects of roughly equal brightness: Cassiopeia A, Cygnus A and the Sun. The first two of these are a supernova remnant and a radio galaxy respectively and are more or less constant in their output. However, on occasions the Sun can brighten by a factor of 10,000, and dominate the radio sky just as it does the optical sky.

It is therefore possible to observe the Sun at radio wavelengths using comparatively simple equipment. Most radio telescopes, even the largest ones, operate on the same basic superheterodyne principles as the ubiquitous transistor radio. In this the radio signal is converted into an electrical signal by the aerial, and this electrical signal is then mixed with the signal from an artificial source oscillating at a fixed frequency. The artificial source is called the local oscillator, and its frequency is close to but different from that received by the radio telescope. The mixed signal contains a component whose frequency is the difference between the input signal frequency and that of the local oscillator. This intermediate frequency is then amplified before being converted to a voltage whose amplitude is proportional to the power in the incoming radio signal.

For detecting the radio Sun, it is entirely feasible for an electronics enthusiast to build his or her own radio telescope. Plans for the receiver may be found from

---

[86] Megahertz or a million cycles per second. The equivalent wavelength is 3 metres.

electronic and radio magazines, and the aerial can be an ordinary TV aerial. The output may simply be fed to a chart recorder or stored in a computer, etc. Care will have to be taken, however, to observe in one of the reserved radio astronomy frequencies,[87] otherwise the solar signal will be drowned by TV transmissions, radio taxis, etc. Alternatively there are a few suppliers that sell ready made receivers for around £500–1000 ($1000–2000).

Observing the radio Sun has none of the safety problems of optical solar observing. The aerial is simply pointed at the Sun, and the signal recorded. However, detection is all that will occur. The beam width of a TV type aerial is 45°, and so the Sun, which is about 1° across at 100 MHz, is completely unresolved. It is possible to improve the resolution[88] of a radio telescope by using a parabolic dish to feed the aerial, and it is these huge dishes which form most people's ideas of what a radio telescope looks like. However, even to start to resolve the Sun at 100 MHz would require a dish over 400 m (1350 feet) across, and so that is scarcely practical.

Professional radio observatories overcome the low resolution of a single radio telescope by combining two telescopes into an interferometer.[89] In an interferometer the signals from the two telescopes are mixed together. Sometimes the signals will be in step with each other (in phase), and they will add to give a stronger total signal. Sometimes the signals will be out of step (180° out of phase), and they will cancel each other out. The resolution of an interferometer is determined by the distance between the two radio telescope, not by their individual sizes. At 100 MHz, that separation would need to be 400 m to start to resolve the Sun.

With two radio telescopes forming an interferometer, it is simple to measure the diameter of the radio Sun. The output from the interferometer will be a series of ups and downs, known as fringes, if it is displayed on a

---

[87] See for example *AT*.

[88] The resolution of an optical telescope or a radio dish is given in degrees by 70 λ/D, where λ is the operating wavelength, and D is the diameter of the telescope objective or radio dish.

[89] The resolution of an interferometer for two point sources is given in degrees by 30 λ/S, where λ is the operating wavelength, and S is the separation of the telescopes. However, for an extended source like the Sun it reverts to 70 λ/S.

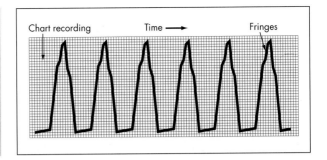

Chart recording   Time ⟶   Fringes

Figure 9.1. The schematic fringe output from a radio interferometer.

chart recorder (Fig. 9.1). The fringes arise because as the Sun moves across the sky, the two signals will change from being in phase to 180° out of phase and back again; alternately reinforcing and then cancelling each other. As the radio telescopes are moved further apart, the fringes will become narrower, but less distinct. Eventually a separation of the telescopes will be reached where the interferometer output is constant. That separation, S, is related to the angular diameter of the Sun in degrees[90] at that wavelength, α, by

$$\alpha = 70\lambda/S.$$

Going further than this, and producing a radio image of the Sun requires large computers and knowledge of some sophisticated mathematics. The process is known as aperture synthesis, and details can be found for example in sources in Appendix 1.

Although obtaining radio images of the Sun is not practical for most amateur astronomers, there is still a good deal of interest to be found in the integrated output from the Sun. The Sun is normally of a similar brightness (depending upon the frequency at which the observations are being made) to Cas A and Cyg A. But this is when it is quiescent. The radio emission is then simply due to the high temperature of the corona (around 1,000,000 K). This is known as thermal emission and the Sun is called the quiet Sun when only thermal radio emission is coming from it. On other occasions however, the Sun can be a thousand or even ten thousand times brighter than this. The extra radiation results from activity such as sunspots and flares (Chapter 5) occurring on the surface of the Sun. The short-wave radio emission (around 3 GHz,

[90] The solar diameter ranges from 0.5° at optical, infrared and microwave frequencies, to 0.75° at 600 MHz and 1° at 60 MHz.

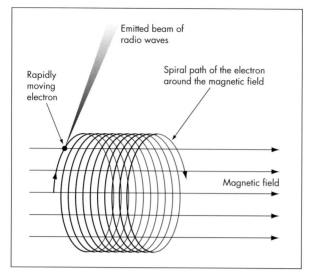

Figure 9.2.
Synchrotron radiation.

100 mm) from the Sun correlates well with the level of solar activity (see Wolf sunspot number, R, Chapter 3) and can be used to monitor it without the weather problems of visual observations of the activity.

The emission from the active Sun is largely due to quite different processes from that of the quiet Sun. It arises mostly as synchrotron radiation,[91] in which very fast electrons, or sometimes protons, are spiralling around magnetic fields (Fig. 9.2). There is also emission from electric charges oscillating back and forth in the solar atmosphere. Both these effects are set in motion by violent explosions occurring in and around large sunspot groups. These explosions, or flares (Chapter 1), send large numbers of electrons and protons at velocities near the speed of light out through the solar atmosphere. The radio emissions then occur as the bursts of particles interact with solar magnetic fields, or upset the normal electrical neutrality of the solar material.

The radio emission from flares occurs as intense bursts lasting a few minutes or so. There are several types of bursts depending upon how their frequency changes with time (Fig. 9.3). If observing with a radio telescope operating at a single frequency, then the bursts will be observed as several separate components as their emissions drift through the observing frequency

---

[91] So called because it can be observed within the particle accelerators or synchrotrons, used by atomic physicists.

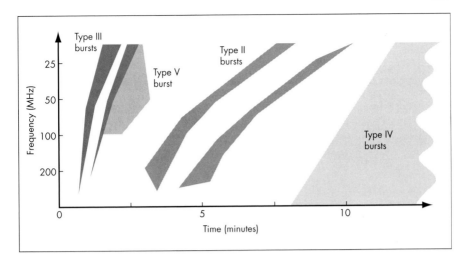

(Fig. 9.4). As well as being of interest in their own right, the radio bursts can be used to alert an optical observer to the occurrence of the flare. These can be easily observed through an Hα filter (Chapter 8), and the most intense flares may sometimes be seen in white light. The particles emitted during a flare and which produce the radio bursts travel out into space and may interact with the Earth's magnetosphere and upper atmosphere to produce aurorae some hours after the flare occurred. So the occurrence of the bursts can also be used as an indicator of possible auroral activity for observers in the right latitudes.[92]

Figure 9.3. Solar radio bursts.

Figure 9.4. Schematic recording of solar radio emission at a single frequency during a set of radio outbursts from a large flare.

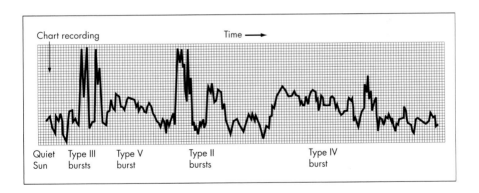

[92] Auroral activity is rarely visible between latitudes 50° S and 50° N.

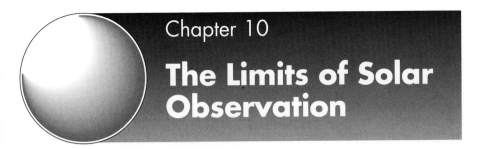

Chapter 10

The Limits of Solar Observation

The Sun is the closest star to us by far, and the only one upon which we can easily see fine detail. It also provides us with vast numbers of photons at all wavelengths. The astronomer's usual problem of picking out the few desired photons from the welter of background photons from other sources is thus often replaced by having to cut down the intensity from the Sun. The abundance of energy and detail has led to many specialised instruments being developed for solar work which cannot be used in normal astronomy. We have seen some of these (Chapter 8): coronagraphs, Hα filters, spectrohelioscopes, prominence spectroscopes, etc., which are within the reach of amateur astronomers, even if in some cases it is at the outer limits of the bounds of possibility. There are many more specialised instruments which are the province purely of the professional astronomer either through cost, the background equipment and knowledge that is needed to operate them and analyse their data, the need for a remote site or a large support team, etc. Details of some of these instruments may be found for example in the author's *AT*. Here we look briefly at three examples to see just what can be done at the limits of solar observing.

Magnetic Fields

Magnetic fields abound on the Sun, ranging from the general magnetic field, which is little stronger than that of the Earth, to the fields in complex, large sunspot groups which may reach 10,000 times the strength of the Earth's magnetic field. The presence of those magnetic fields is revealed through a detailed analysis of the solar spectrum. This is possible because a spectrum line originating in the presence of a magnetic field is split into several components. The phenomenon is known as the Zeeman effect.[93] In simple cases the line splits into two components when the magnetic field is along the line of sight, and into three components when it is perpendicular to the line of sight. Furthermore the polarisation of the components is circular[94] in the first case and plane in the second, so that it is possible to separate out the individual effects even when the line of sight is at some intermediate angle to the magnetic field. There are several designs of magnetometer for solar work,[95] but most work by passing the light of a spectrum line through a polariser to separate out one of the Zeeman components, and then switching the polariser to allow the other component through. By comparing the output in the two cases, magnetic fields as weak as one tenth of the Earth's magnetic field can be detected on the Sun.

Spacecraft

Numerous spacecraft have been launched specifically to study the Sun, and many more have carried instruments to observe the Sun along with other items in their payloads. Although few people even amongst the ranks of the professional astronomers are able to design and build their own instruments to go on spacecraft, many observers can make use of the results.

Highlights from spacecraft missions are now posted at the relevant web sites and home pages for those

---

[93] Strictly the inverse Zeeman effect for absorption lines, for further details see *OAS*.

[94] Linear polarisation of light will be familiar to most people from the ubiquitous polarised sunglasses. This type of polarisation results from the vibration direction of the light being in a fixed direction. In unpolarised light the vibration directions are randomised. In circularly polarised light the vibration direction rotates through a full circle at the frequency of the radiation. For visible light therefore the direction of vibration is rotating some 10^{14} to 10^{15} times per second.

[95] For further details see *AT*.

missions on the internet. They can be found by searching for the spacecraft's name. Frequently images, data, and results appear on web sites long before they are published in even the popular astronomy journals, and probably a year or more before they appear in research journals.

Many space missions operate archives which contain their older data. These archives can be accessed via computer links by approved researchers, and this can include amateur astronomers, teachers and others. A commonly used system is that the original research team will have sole access to the data for six months or a year, after which it is made available in the archive. So much data is produced by space missions, that not all of it can be processed in the time available, and even data that has been looked at for one purpose can often contain useful information for other purposes. The archive data is therefore a very valuable resource that can with relative ease be used by those with a serious interest in solar studies.

Solar Oscillations

The whole Sun is ringing like a bell with periods for its complex vibrations of a few minutes. The changes in the Sun's shape and the resulting velocity patterns on its surface are, however, very tiny. Their study therefore requires very sophisticated techniques. The resonance scattering spectrometer works by passing a solar spectrum line originating from potassium or sodium through a hot vapour of the same element. The solar light is circularly polarised, and a similar polarisation is induced into the vapour's absorption by a magnetic field passing through it. By switching the direction of the magnetic field, the vapour's circular polarisation can be changed rapidly, and so the intensities at the edges of the solar spectrum line compared. If the line is Doppler shifted, the balance of those intensities will change, and so allow the velocity of the part of the Sun being studied to be measured. So sensitive is this device that velocities slower than 1 mm/s can be detected – quite literally a snail's pace!

Solar Image Gallery

A Whole Sun – White Light

A1 22 March 2000. Obtained at the Maple Ridge Observatory using a 60 mm refractor with a telecompressor to give focal ratios in the range f5.75 to f7, a Thousand Oaks® Type II full aperture filter, and a Pixcel® 237 CCD camera. Image coloured using PaintShop Pro 6. Reproduced by courtesy of Brian Colville.

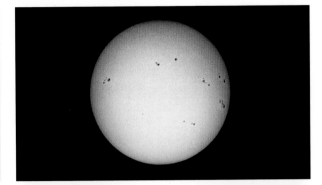

A2 19 March 2000. Obtained using a Meade® ETX telescope with a 25 mm eyepiece and an Orion® full aperture solar filter. The camera, a digital Fuji® FX1500, was hand-held to point through the eyepiece, and the camera's auto systems did the rest. Image enhanced using PaintShop® Pro 6 unsharp masking. Reproduced by courtesy of John Watson.

A3 Plenty of spots and faculae, 29 January 1991. Obtained using a 180 mm refractor with a heat rejection filter. Reproduced by courtesy of Prof. Jean Dragesco.

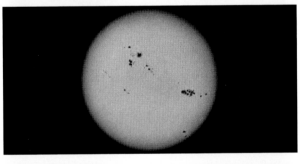

A4 Sunspot groups on 28 February 2000. Photographed at the prime focus of a 90 mm Meade® ETX telescope with a full aperture Thousand Oaks® filter in place. The image was then scanned and the colour adjusted to give a natural appearance. Reproduced by courtesy of John McConnell.

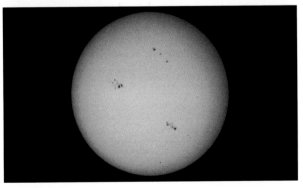

B Solar Details – White Light

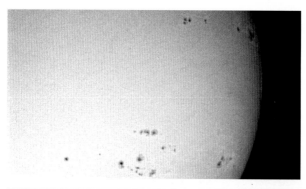

B1 Sunspots and faculae on the western edge of the Sun, 24 June 1999. Obtained using a Celestron® 80 mm refractor with a home-made full aperture filter using BAADER® AstroSolar film and a red filter. The camera is a StarlightXpress® CCD detector and the images have been processed by contrast stretching and unsharp masking. Reproduced by courtesy of Mike Beales.

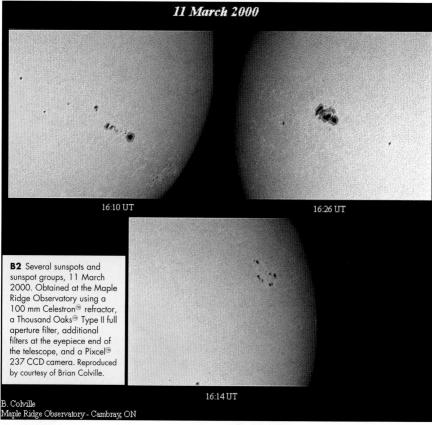

11 March 2000

16:10 UT   16:26 UT

B2 Several sunspots and sunspot groups, 11 March 2000. Obtained at the Maple Ridge Observatory using a 100 mm Celestron® refractor, a Thousand Oaks® Type II full aperture filter, additional filters at the eyepiece end of the telescope, and a Pixcel® 237 CCD camera. Reproduced by courtesy of Brian Colville.

16:14 UT

B. Colville
Maple Ridge Observatory - Cambray, ON

B3 Solar granulation imaged at a wavelength of 1.56 μm in the near infrared by the 0.65 m vacuum reflector at the Big Bear Solar Observatory on 12 March 1999. The field of view is 65″ by 50″. Reproduced by courtesy of the Big Bear Solar Observatory and the New Jersey Institute of Technology.

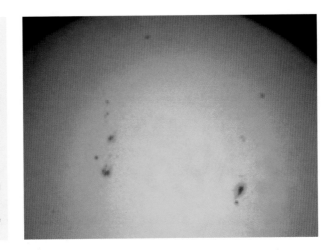

B4 Sunspots on 13 February 1999. Obtained by Mike Weasner using a Meade® ETX telescope with a Solar® II type 2 plus full aperture solar filter and a wide field adaptor. Reproduced by courtesy of Mike Weasner.

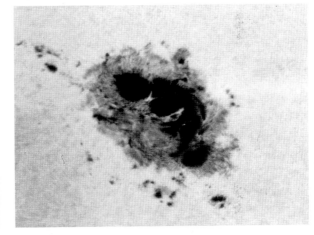

B5 A large sunspot group on 9 June 1991. Obtained using a 180 mm refractor with a heat rejection filter. Reproduced by courtesy of Prof. Jean Dragesco.

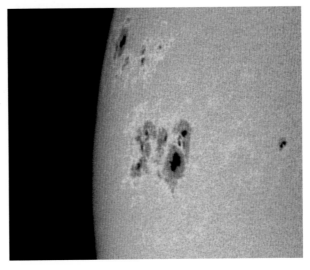

B6 A sunspot group and faculae near the solar limb and showing the Wilson effect, 13 May 2000. Obtained using a 200 mm Meade® LX200 telescope stopped down to 60 mm, a Thousand Oaks® full aperture filter and a density 8 neutral density filter. A 0.3 s exposure on an HX516 CCD camera. Reproduced by courtesy of Peter Garbett.

C Whole Sun – Narrow Band

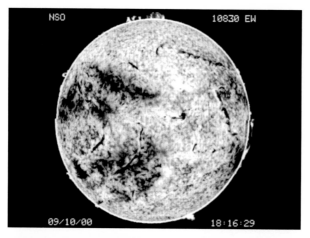

C1 An image of the Sun in the light of ionised helium at 1083.0 nm. The image shows only the chromosphere and is based upon the strength (equivalent width) of the spectrum line. Obtained on 10 September 2000 using the vacuum telescope of the National Solar Observatory at Kitt Peak. Reproduced by courtesy of the National Solar Observatory.

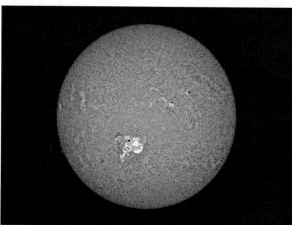

C2 Hα image showing a large active region with many plages, 19 July 2000. Obtained using a 70 mm Pronto® refractor with a 0.33× focal reducer, a Daystar® 0.06 nm Hα filter and recorded with a Hi-SIS 43 CCD camera. Reproduced by courtesy of Thierry Legault.

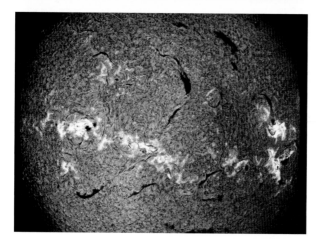

C3 Hα image showing a great deal of activity with many plages and filaments, 2 March 1992. Reproduced by courtesy of Prof. F. Rouvière.

D Solar Details – Narrow Band

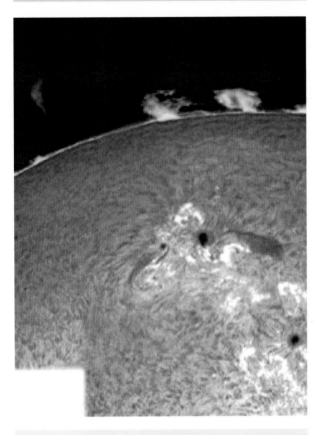

D1 A solar active region in Hα, showing sunspots, filaments, plages and prominences. Obtained using a Coronado® Hα filter on an Astrovid® 2000 refractor with a 3× Barlow lens on 12 August 2000. The image is a composite of five raw images processed using Photoshop® 5.0 and PictureWindows® 2.5. Reproduced by courtesy of Dr Fritz Hemmerich.

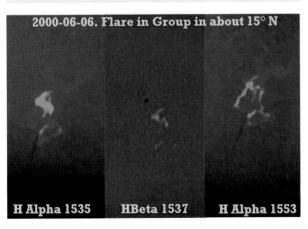

2000-06-06. Flare in Group in about 15° N

H Alpha 1535   HBeta 1537   H Alpha 1553

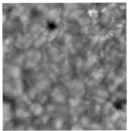

D2 Solar granulation at a wavelength of 430.1 nm with improved resolution being obtained by combining 200 images using speckle masking. The field of view is 15.5" by 15". Obtained at the Big Bear Solar Observatory on 14 August 1997. Reproduced by courtesy of the Big Bear Solar Observatory and the New Jersey Institute of Technology.

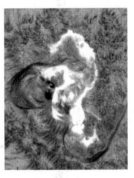

D3 The giant solar flare (known as the Sea Horse flare) of 7 August 1972. Obtained at the Big Bear Solar Observatory. Reproduced by courtesy of the Big Bear Solar Observatory and the New Jersey Institute of Technology.

D4 A solar flare at Hα and Hβ wavelengths, 6 June 2000. Obtained using a home-made spectrohelioscope and processed using PaintShop® Pro 5. A plot of the radio emission at 136 MHz is shown in image E6. Reproduced by courtesy of Commander Henry Hatfield.

E Prominences

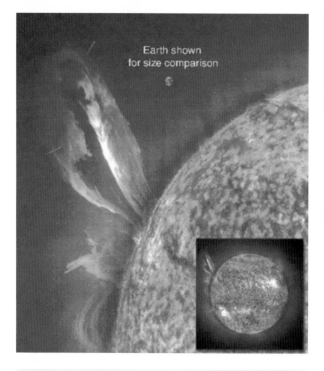

Earth shown
for size comparison

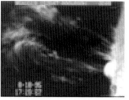

E2 An eruptive flare and prominence on the solar east limb. Obtained at the Big Bear Solar Observatory on 18 August 1995. Reproduced by courtesy of the Big Bear Solar Observatory and the New Jersey Institute of Technology.

E1 A large eruptive prominence seen in the light of ionised helium at a wavelength of 30.4 nm on 24 July 1999. The Earth has been added to give the scale of the event. Obtained by the SOHO (SOlar and Heliospheric Observatory) spacecraft. Reproduced by courtesy of the SOHO/EIT Consortium NASA/ESA.

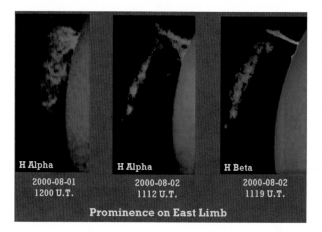

H Alpha

H Alpha

H Beta

2000-08-01
1200 U.T.

2000-08-02
1112 U.T.

2000-08-02
1119 U.T.

Prominence on East Limb

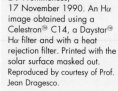

E4 Prominences, 17 November 1990. An Hα image obtained using a Celestron® C14, a Daystar® Hα filter and with a heat rejection filter. Printed with the solar surface masked out. Reproduced by courtesy of Prof. Jean Dragesco.

E3 Development of prominences on the Sun's east limb, 2 August 2000. Hα and Hβ images obtained using a home-made spectro-helioscope and processed using PaintShop® Pro 5. Reproduced by courtesy of Commander Henry Hatfield.

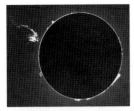

E5 A spectacular prominence on 20 March 2000. Obtained using a 100 mm coronagraph. Reproduced by courtesy of Daniel Lachaud.

F Eclipses

F1 Baily's beads during the total eclipse on 26 February 1998. Obtained at Aruba on Fujichrome® Sensia 100 slide film using a Questar® 88 mm telescope directly attached to an Olympus® OM2S camera body and with a full aperture filter. Reproduced by courtesy of Jack Morris.

F2 The diamond ring effect during the total eclipse on 26 February 1998. Obtained at Aruba on Fujichrome® Sensia 100 slide film using a Questar® 88 mm telescope directly attached to an Olympus® OM2S camera body and with a full aperture filter. Reproduced by courtesy of Jack Morris.

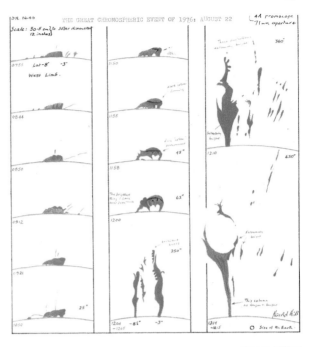

E6 The development of a spectacular prominence over a period of 4 hours on 22 August 1976. A series of drawings made using a prominence spectroscope on a 71 mm telescope. Reproduced by courtesy of Harold Hill.

G Coronae

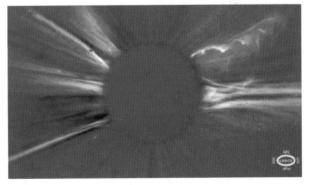

G1 An image of coronal streamers extending to more than 3 000 000 km above the Sun from the SOHO (SOlar and Heliospheric Observatory) spacecraft. Obtained using the LASCO (Large Angle Spectroscopic COronagraph) instrument on 21 August 1996. Reproduced by courtesy of the SOHO/LASCO Consortium NASA/ESA.

H Radio, X-ray and Specialised Imaging

I Instruments

H1 A magnetogram showing the north and south magnetic fields on the Sun (red and blue), and areas of neutral or low magnetic field strength (white). Obtained at 10.00 am on 10 September 2000 using the magnetograph of the 150-foot solar tower telescope of the Mount Wilson Observatory. The solar tower telescope is operated by UCLA with funding from NASA, ONR and NSF, under agreement with the Mount Wilson Institute. Reproduced by courtesy of the Mount Wilson Institute.

I1 A Questar® 88 mm telescope with a Coronado® 0.6ATM Hα solar filter and sun shield in place. Reproduced by courtesy of the Questar Corporation.

H2 A dopplergram showing the velocities of material in the solar surface layers towards or away from the Earth. Obtained at 10.00 am on 10 September 2000 using the magnetograph of the 150-foot solar tower telescope of the Mount Wilson Observatory. The solar tower telescope is operated by UCLA with funding from NASA, ONR and NSF, under agreement with the Mount Wilson Institute. Reproduced by courtesy of the Mount Wilson Institute.

I2 A Meade® ETX telescope with an Orion® full aperture solar filter in place. Reproduced by courtesy of John Watson.

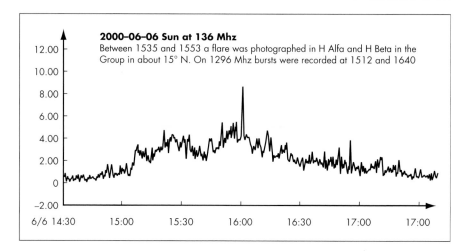

2000–06–06 Sun at 136 Mhz
Between 1535 and 1553 a flare was photographed in H Alfa and H Beta in the Group in about 15° N. On 1296 Mhz bursts were recorded at 1512 and 1640

H3 A plot of the radio emission from the flare on 6 June 2000 (see image D4) at a frequency of 136 MHz. Obtained using a home-made Yagi aerial driven to follow the Sun with the noise level being logged every 2 seconds. Reproduced by courtesy of Commander Henry Hatfield.

Appendix 1

Bibliography

Magazines and Journals

Only the major and relatively widely available journals are listed. There are numerous more specialised research level journals available in academic libraries.

Popular

Astronomy
Astronomy Now
Ciel et Espace
Journal of the British Astronomical Association
New Scientist
Publications of the Astronomical Society of the Pacific
Scientific American
Sky and Telescope

Research

Astronomical Journal
Astronomy and Astrophysics
Astrophysical Journal
Monthly Notices of the Royal Astronomical Association
Nature
Science
Solar Physics

Books

Abbreviated References

Astrophysical Techniques [AT], C.R. Kitchin, 3rd edition, IoP Press, 1998.
Optical Astronomical Spectroscopy [OAS], C.R. Kitchin, IoP Press, 1995.
Telescopes and Techniques [T&T], C.R. Kitchin, Springer-Verlag, 1995.

The Sun and Solar Eclipses

Astrophysics of the Sun, H. Zirin, Cambridge University Press, 1988.
Cambridge Eclipse Photography Guide, J.M. Pasachoff, M.A. Covington, Cambridge University Press, 1992.
Canon of Solar Eclipses, J. Meeus, C.C. Grosjean, W. Vanderleen, Pergamon Press, 1966.
Discovering the Secrets of the Sun, R. Kippenhahn, John Wiley, 1994.
Guide to the Solar Corona, D.E. Billings, Academic Press, 1966.
Observing the Sun, P.O. Taylor, Cambridge University Press, 1991.
Physics of the Solar Corona, I.S. Shklovskii, Pergamon Press, 1965.
Quiet Sun, E.G. Gibson, NASA, 1973.
Solar Astrophysics, P.V. Foukal, John Wiley, 1990.
Solar Chromosphere, R.J. Bray, R.E. Loughhead, Chapman and Hall, 1974.
Solar Granulation, R.J. Bray, R.E. Loughhead, C.J. Durrant, Cambridge University Press, 1984.
Solar Prominences, E. Tandberg-Hanssen, D. Reidel, 1974.
Solar System Astrophysics, J.C. Brandt, P.W. Hodge, McGraw-Hill, 1964.
Sun, G. Abetti, Faber and Faber, 1957.
Sun, M. Stix, Springer-Verlag, 1991.
Sun, Earth and Sky, K. Lang, Springer-Verlag, 1995.
Sun in Eclipse, M. Maunder, P. Moore, Springer-Verlag, 1998.
Sunspots, R.J. Bray, R.E. Loughhead, Chapman and Hall 1964.
Totality–Eclipses of the Sun, M. Littmann, K. Willcox, University of Hawaii Press, 1991.

Practical Astronomy Books

Amateur Astronomer's Handbook, J.B. Sidgwick, Faber and Faber, 1971.

Astronomical Telescope, B.V. Barlow, Wykeham Publications, 1975.
Astrophysical Techniques, C.R. Kitchin, 3rd edition, IoP Press, 1998.
Beginners Guide to Astronomical Telescope Making, J. Muirden, Pelham Press, 1975.
Building and Using an Astronomical Observatory, P. Doherty, Stevens, 1986.
Eyes on the Universe, P. Moore, Springer-Verlag, 1997.
Manual of Advanced Celestial Photography, B.D. Wallis, R.W. Provin, Cambridge University Press, 1988.
Optical Astronomical Spectroscopy, C. Kitchin, IoP Press, 1995.
Seeing Stars, C. Kitchin, R.W. Forrest, Springer-Verlag, 1998.
Solar System: A Practical Guide, D.Reidy, K. Wallace, Allen and Unwin, 1991.
Star Gazing through Binoculars: A Complete Guide to Binocular Astronomy, S. Mensing, TAB, 1986.
Star Hopping: Your Visa to the Universe, R.A. Garfinkle, Cambridge University Press, 1993.
Telescopes and Techniques, C. Kitchin, Springer-Verlag, 1995.

Introductory Books

Astronomy: A Self-Teaching Guide, D.L. Moche, John Wiley, 1993.
Astronomy: The Evolving Universe, M. Zeilik, John Wiley, 1994.
Astronomy: Principles and Practice, A.E. Roy, D. Clark, Adam Hilger, 1988.
Astronomy through Space and Time, S. Engelbrektson, WCB, 1994.
Introductory Astronomy and Astrophysics, M. Zeilik, S.A. Gregory, E.v.P. Smith, Saunders, 1992.
Universe, W.J. Kaufmann III, W.H. Freeman, 1994.

Catalogues, Atlases and Reference Books

Astronomical Almanac (published for each year), HMSO/US Government Printing Office.
Astrophysical Quantities, C.W. Allen, Athlone Press, 1973.
Atlas of Representative Stellar Spectra, Y. Yamashita, K. Nariai, Y. Norimoto, University of Tokyo Press, 1977.
Handbook of the British Astronomical Association (published for each year), British Astronomical Association.
Messier's Nebulae and Star Clusters, K.G. Jones, Cambridge University Press, 1991.
Norton's 2000.0, I. Ridpath (ed.), Longman, 1989.

Sky Atlas 2000.0, W. Tirion, Sky Publishing Corporation, 1981.
Sky Catalogue 2000, volumes 1 and 2, A. Hirshfield, R. W. Sinnott, Cambridge University Press, 1985.

Appendix 2

Equipment Suppliers

NB Inclusion of a supplier or manufacturer in the list below does not constitute a recommendation or endorsement by the author or publishers of the supplier's products. Readers are advised to obtain full information and competitive quotes before making any purchases.

Solar Filters

Abel Express
KDKK Industrial Complex, Building 2
100 Rosslyn Road
Carnegie
PA 15106
USA

Astro-Physics
11250 Forest Hills Road
Rockford
IL 61111
USA

Astronomical Innovations
PO Box 14853
Lenexa
KS 66285
USA

Broadhurst Clarkson and Fuller
Telescope House
63 Farringdon Road
London EC1M 3JB
United Kingdom

David Hinds
Unit 34
The Solk Mill
Brook Street
Tring HP 23 5 EE
United Kingdom

Day Star Filters
PO Box 5110
Diamond Bar
CA 91765
USA

Eclipse Ltd.
Belle Etoile
Rue du Hamel
Castel
Guernsey GY 5 7QJ
United Kingdom

Lumicon
2111 Research Drive
#4 Livermore
CA 94550
USA

Roger W. Tuthill Inc.
Box 1086
11 Tanglewood Lane
Mountainside
NJ 07092-0086
USA

Thousand Oaks Optical
Box 4813
Thousand Oaks
CA 91359
USA

Telescopes and Accessories

AE Optics Ltd.
Vega Court
East Drive
Caldecot
Cambridge CB3 7NZ
United Kingdom

Astro-physics Inc.
11250 Forest Hills Road
Rockford
IL 6115
USA

Broadhurst Clarkson and Fuller
Telescope House
63 Farringdon Road
London EC1M 3JB
United Kingdom

Carl Zeiss Jena GmbH
Postfach 125
Tatzendpromenade 1A
Jena O-6900
Germany

Celestron Telescopes,
PO Box 3578
Torrance
CA 90503
USA

Dark Star Telescopes
6 Pinewood Drive
Ashley Heath
Market Drayton
Shropshire TF9 4PA
United Kingdom

Meade Instrument Co.
16542 Millikan Avenue
Irvine
CA 93714
USA

Questar Corp.
Route 202, Box 59
New Hope
PA 18938
USA

Scope City
71 Bold Street
Liverpool L1 4EZ
United Kingdom

Torus Optical
67 Bon-Aire
Iowa City
IA 52240
USA

Unitron
170 Wilbur Place
PO Box 469
Bohemia
NY 11716
USA

Appendix 3

Web Sites Relating to the Sun

NB This is a representative sample of available sites only. Searches for combinations of Sun / Solar / Sol / Eclipses / Photographs / Images / Data / Telescopes / Spacecraft, etc., will soon reveal many more sites.

http://america.net/~boo/html/sun_gun.html — Sun Gun (home made solar telescope)

http://argo.tuc.noao.edu/nsokp/mainlinkn3.html — National Solar Observatory

http://bolero.gsfc.nasa.gov/~batchelo/daveb.html — Dave Batchelor's solar image page

http://fusedweb.pppl.gov/CPEP/Chart_pages/5.Plasmas/-SunLayers.html — From Core to Corona

http://lucille.physics.uwo.ca/SunEarth/ — Sun–Earth connection

http://outworld.compuserve.com/homepages/Larry_Freeman/-dials.htm — Larry Freeman's Sun Dials

http://pcsinspace.hst.nasa.gov/space/sun.htm — Exploring the Sun

http://sohowww.nascom.nasa.gov/ — Solar and Heliospheric Observatory

http://solar-center.stanford.edu/eclipse/eclipse.html — Stanford Solar Center's eclipses

http://solar.physics.montana.edu/YPOP/Classroom/Lessons/-Filters/sunfilters.html — Many Faces of the Sun

http://solar.uleth.ca/sunnow — Current Solar Halpha image

http://sunmil1.uml.edu/eyes/ — Robotic Solar Observatory

http://tlc.discovery.com/tlcpages/sunstorms/-sunstorms_main.html — Sunstorms Learning Channel

http://ulysses.jpl.nasa.gov/science/results.html — Ulysses spacecraft

http://umbra.gsfc.nasa.gov/ — Solar Data Analysis Center

http://umbra.nascom.nasa.gov/eclipse/ — Solar Data Analysis Center Eclipse information

http://windows.engin.umich.edu/sparc/rtdata/sun.html — Sun Today (University of Michigan)

http://www.astro.ucla.edu/~obs/intro.html — Mount Wilson 150-foot solar telescope site – a very good site with links to the following additional sites:
  AAS Solar Physics Division
  Base Solaire Sol 2000
  Big Bear Solar Observatory
  BISON: Birmingham Solar Oscillations Network

Centre de prévision de l'activité
solaire et géomagnetique RWC
Paris-Meudon
Culgoora Solar Observatory
Exploratorium's "Solar Max 2000"
GOLF: Global Oscillations at Low
Frequencies
GONG: Global Oscillations Network
Group
High Altitude Observatory
Great moments in the history of
solar physics
Hiraiso Solar Terrestrial Research
Center
Kanzelhöhe Solar Observatory
Kiepenheuer-Institute für
Sonnenphysik
Learmonth Solar Observatory
Lockheed–Martin Solar and
Astrophysics Laboratory
Majestic research
Marshall Space Flight Centre Solar
Physics Branch
Mauna Loa Solar Observatory
Mees Solar Observatory
Mount Wilson Observatory 60-foot
Solar Tower
National Solar Observatory, Kitt
Peak
National Solar Observatory,
Sacrament Peak
NGDC Solar and Upper
Atmospheric Data Services
Nobeyama Solar Radio Telescope
San Fernando Observatory
SDAC: Solar Data Analysis Centre
SHINE
Siberian Solar Radio Telescope
SOHO: Solar and Heliospheric
Observatory
SOI: Solar Oscillations Investigation
Solar Ephemeris
Solar Observing by Peter Meadows
Solar Terrestrial Activity Report
SOLIS: Synoptic Optical Long-Term
Investigations of the Sun
Space Environment Center
SIDC: Sunspot Index Data Centre
Swedish Solar Telescope at La
Palma
Transition Region and Coronal
Explorer
VIRGO: Variability of Solar
Irradiance and Gravity Oscillations
Wilcox Solar Observatory
Yohkoh SXT Spacecraft

http://www.eso.org/	European Southern Observatory
http://www.ast.cam.ac.uk/~baa/baamain.html	British Astronomical Association
http://www.bbso.njit.edu/	Big Bear Solar Observatory
http://www.cassfos01.ucsd.edu/public/tutorial/Sun.html	University of California, San Diego
http://www.ct.astro.it/sunoac.html	Catania Astrophysical Observatory – today's solar image
http://www.dasop.obspm.fr/previ/pagetipe.htm	Page de liens. A reference site with links to many of the sites given by the Mount Wilson 150-foot solar telescope site plus:

    The Dynamic Sun
    Les techniques d'observation du
      soleil
    Beginner's Guide to the Sun
    Solar Guide
    The Sun
    Solar–Terrestrial Physics Archives
    Physique solaire
    Solar Physics links
    Solar Links
    Solar observatories, institutes and
      data centers links
    Laboratoire de physique du soleil et
      de l'héliosphère
    Catania Solar Observatory
    Radiohéliographe de Nançay
    Solar Radio Flux at 2800 MHz,
      Penticton
    ESA Space weather
    Primer on the Solar Space
      Environment
    Space Weather: NRC
    Space Science Institute, Boulder
    Space Physics Data System
    Current Space Weather Status
    Latest 48 hours of solar wind data
    Ace Real Time Solar Wind
    Le vent solaire et le champ
      magnétique interplanétaire
    Physique spatial
    Centre de prévision solaire
    Sunspot index data center, Brussels

http://www.discovery.com/cams/sun/sun.html	Discovery Space Cameras
http://www.hao.ucar.edu/public/slides/slides.html	Pictorial Introduction to the Sun
http://www.hawastsoc.org/solar/cap/sun/sun.htm	Prominences
http://www.hawastsoc.org/solar/cap/sun/sunspot.htm	Sunspots
http://www.ips.oz.au/culgoora/	Culgoora Solar Observatory
http://www.lhp.ac.cn/	Laboratory of Numerical Study of Heliospheric Physics
http://www.pvamu.edu/cps/solar.html	Prairie View Solar Observatory
http://www.quicklinks.on.ca/~maple/index.html	Maple Ridge Observatory
http://www.sciam.com/exhibit/111698sun/	Simulating Sol
http://www.sec.noaa.gov/current_images.html	Solar Web Site index – Space Environment Center
http://www.sel.noaa.gov/images/gif/9901_full/-99012502431h3o.gif	Learmonth Observatory

http://www.sel.noaa.gov/solar_sites.html	Solar Site Index
http://www.skypub.com/index.shtml	*Sky and Telescope* magazine
http://www.spaceart.com/solar/eng/sun.htm	Space Art – The Sun
http://www.-ssi.colorado.edu/ExploringSpace/VirtualExhibit/-TheDynamicSun/1.html	Space Science Institute – Dynamic Sun
http://www.ultisoft.demon.co.uk/ecring.html	Eclipse chasers web ring
http://www.weasner.com/etx/menu.html	Mike Weasner's ETX image collection
http://wwwssl.msfc.nasa.gov/ssl/pad/solar/sunactv.htm	Sun in Action
http://wwwssl.msfc.nasa.gov/ssl/pad/solar/whysolar.htm	NASA – Why we Study the Sun

Index

NB – an 'm' after a page number indicates the start of a major section on the topic, a 't' after the page number indicates a table of values for that topic.